SAFE ABOVEGROUND STORAGE TANKS

HEARING

BEFORE THE

SUBCOMMITTEE ON
TRANSPORTATION AND HAZARDOUS MATERIALS

OF THE

COMMITTEE ON
ENERGY AND COMMERCE
HOUSE OF REPRESENTATIVES

ONE HUNDRED THIRD CONGRESS

SECOND SESSION

SEPTEMBER 14, 1994

Serial No. 103–161

Printed for the use of the Committee on Energy and Commerce

U.S. GOVERNMENT PRINTING OFFICE

86469CC

WASHINGTON : 1995

For sale by the U.S. Government Printing Office
Superintendent of Documents, Congressional Sales Office, Washington, DC 20402
ISBN 0-16-046804-3

COMMITTEE ON ENERGY AND COMMERCE

JOHN D. DINGELL, Michigan, *Chairman*

HENRY A. WAXMAN, California
PHILIP R. SHARP, Indiana
EDWARD J. MARKEY, Massachusetts
AL SWIFT, Washington
CARDISS COLLINS, Illinois
MIKE SYNAR, Oklahoma
W.J. "BILLY" TAUZIN, Louisiana
RON WYDEN, Oregon
RALPH M. HALL, Texas
BILL RICHARDSON, New Mexico
JIM SLATTERY, Kansas
JOHN BRYANT, Texas
RICK BOUCHER, Virginia
JIM COOPER, Tennessee
J. ROY ROWLAND, Georgia
THOMAS J. MANTON, New York
EDOLPHUS TOWNS, New York
GERRY E. STUDDS, Massachusetts
RICHARD H. LEHMAN, California
FRANK PALLONE, JR., New Jersey
CRAIG A. WASHINGTON, Texas
LYNN SCHENK, California
SHERROD BROWN, Ohio
MIKE KREIDLER, Washington
MARJORIE MARGOLIES-MEZVINSKY,
  Pennsylvania
BLANCHE M. LAMBERT, Arkansas

CARLOS J. MOORHEAD, California
THOMAS J. BLILEY, JR., Virginia
JACK FIELDS, Texas
MICHAEL G. OXLEY, Ohio
MICHAEL BILIRAKIS, Florida
DAN SCHAEFER, Colorado
JOE BARTON, Texas
ALEX McMILLAN, North Carolina
J. DENNIS HASTERT, Illinois
FRED UPTON, Michigan
CLIFF STEARNS, Florida
BILL PAXON, New York
PAUL E. GILLMOR, Ohio
SCOTT KLUG, Wisconsin
GARY A. FRANKS, Connecticut
JAMES C. GREENWOOD, Pennsylvania
MICHAEL D. CRAPO, Idaho

ALAN J. ROTH, *Staff Director and Chief Counsel*
DENNIS B. FITZGIBBONS, *Deputy Staff Director*
MARGARET A. DURBIN, *Minority Chief Counsel and Staff Director*

SUBCOMMITTEE ON TRANSPORTATION AND HAZARDOUS MATERIALS

AL SWIFT, Washington, *Chairman*

BLANCHE M. LAMBERT, Arkansas
W.J. "BILLY" TAUZIN, Louisiana
RICK BOUCHER, Virginia
J. ROY ROWLAND, Georgia
THOMAS J. MANTON, New York
GERRY E. STUDDS, Massachusetts
FRANK PALLONE, JR., New Jersey
LYNN SCHENK, California
PHILIP R. SHARP, Indiana
EDWARD J. MARKEY, Massachusetts
BILL RICHARDSON, New Mexico
JOHN D. DINGELL, Michigan
  (Ex Officio)

MICHAEL G. OXLEY, Ohio
JACK FIELDS, Texas
DAN SCHAEFER, Colorado
FRED UPTON, Michigan
BILL PAXON, New York
PAUL E. GILLMOR, Ohio
MICHAEL D. CRAPO, Idaho
CARLOS J. MOORHEAD, California
  (Ex Officio)

ARTHUR P. ENDRES, JR., *Staff Director/Chief Counsel*
KRISTINA M. LARSEN, *Staff Assistant*

(II)

CONTENTS

(III)

SAFE ABOVEGROUND STORAGE TANKS

WEDNESDAY, SEPTEMBER 14, 1994

HOUSE OF REPRESENTATIVES,
COMMITTEE ON ENERGY AND COMMERCE,
SUBCOMMITTEE ON TRANSPORTATION
AND HAZARDOUS MATERIALS,
Washington, DC.

The subcommittee met, pursuant to notice, at 10 a.m., in room 2322, Rayburn House Office Building, Hon. Al Swift (chairman) presiding.

Mr. SWIFT. The subcommittee will come to order.

I would like to welcome everyone to the hearing on H.R. 1360, the Safe Aboveground Storage Tank Act of 1993.

This bill provides an answer to what some see as a lack of comprehensive regulation of aboveground storage tanks. Unlike underground storage tanks that are regulated by EPA under Title I of the Resource Recovery Act, aboveground storage tanks are not regulated under one comprehensive law.

The regulation of aboveground storage tanks fall under more than five different statutes. A principle question in this hearing is whether this multi-statute approach provides a comprehensive level of protection to human health and the environment or whether it leaves significant gaps that can result in the release of hazardous chemicals.

There have been a number of highly publicized incidents since the mid-1980's involving releases of aboveground storage tanks. In some cases, they have ruptured, spilling large amounts of their contents in a short period of time.

A good example is a 1988 failure of a tank in this wonderful town in Pennsylvania that spilled approximately a million gallons of oil in the Monongahela River. I can pronounce Monongahela, but I was stymied by Floreffe, Pennsylvania.

In other instances, small leaks in aboveground tanks have over time contaminated groundwater and spread beyond the tank facility boundaries. One of these, a leak discovered in 1990 at a petroleum facility in Fairfax, Virginia, resulted in the contamination of dozens of neighboring homes.

Today, we are fortunate to have an expert on that incident and the author of this legislation, Representative Jim Moran, to help us better understand this problem and the solution offered by H.R. 1360.

I look forward to Congressman Moran's statement and the testimony of all our witnesses, and I am happy to recognize the ranking Republican in the subcommittee, the gentleman from Ohio.

(1)

Mr. OXLEY. Thank you, Mr. Chairman. I, too, would like to welcome our colleague, Jim Moran. At today's hearing, we are using the legislative proposal for the further expansion of Federal regulatory programs to be administered by EPA. This Congress, we have all examined some of EPA's existing programs, such as Superfund and the Safe Drinking Water Act and the Municipal Landfill Program, and found them to be very problematic.

Before we entertain further expansion, we must seriously look at Congress' and EPA's ability to fix the problems in existing Federal programs, something neither Congress nor EPA has yet to accomplish.

I am pleased to begin to lay out a legislative record in the area of aboveground storage tanks. Indeed, I could not support further expansion of EPA authority without, first, having a strong and reliable statement of the problem warranting Federal attention and, second, a clear indication of why State and local mechanisms routinely fail.

The RCRA corrective action program is a disaster, just as Superfund is a disaster. Does Washington know the best answers in the area of aboveground tanks?

I understand that the American National Standards Institute and the American Petroleum Institute, in consultation with interested parties, have put together standards for the design, construction, operation, maintenance and inspection of terminal and tank facilities. This constitutes an industry standard of care that can be used in State or local programs with reliability purposes. Is there a debate about these standards and what new expertise will the EPA bring to this issue?

At our last hearing, we looked at the problem of one-size-fits-all requirements for groundwater monitoring for landfills in Texas and Alaska. Are all storage tank operations really the same in Louisiana or California?

These are the kinds of critical questions that I would need resolved before seriously entertaining support for Federal legislation in this area.

And, finally, as noted in EPA's testimony, any legislation must provide for a risk-based approach so that both public and private sector resources can be directed towards the most serious threats.

Mr. Chairman, I look forward to hearing from today's witnesses to begin to answer these questions.

Mr. SWIFT. I thank the gentleman and am happy to welcome a gentleman, member of the full committee, who is quickly becoming a de facto member of this subcommittee, Alex McMillan, for an opening statement.

Mr. MCMILLAN. I thank the Chair for allowing me to join the subcommittee. It seems that most of my local problems are under the jurisdiction of this subcommittee, so I appreciate your consideration.

I, like Congressman Moran, became aware of the problem of leaking of aboveground storage tanks when constituents in the Paw Creek community, as it is known, west of the city of Charlotte, located near a major tank farm, began to complain of increased cancer rates in the community.

This tank farm, I suppose like many, had really begun to develop in the 1940's and was a major terminal on the pipeline in the area, so there is quite a bit of oil storage capacity there, much of it put in at a very different time, with a different attitude towards these things. Studies of local wells indicated substantial amounts of contamination but didn't seem to be able to fix responsibility. Some of the stories that came out of that were rather shocking, having to do with the amount of petroleum floating on top of the water table and so on. I am not going to try to get into all of those problems, regardless of what was said, the pollution was considerable.

The community logically turned to the petroleum companies involved, and there were a number of them, for assistance and to establish some sort of cleanup schedule, and I would have to say that in the first instance, from their perspective, they were stonewalled. I think part of that arose out of confusion as to who was actually the cause of the problem.

The fact of the matter is that because the tank farm was there, the problem was there, and I think the response was extremely slow to begin with. The efforts to operate through State regulation were certainly, to the people who were affected, extremely frustrating and slow, although the State of North Carolina is pretty good in most instances. But, nevertheless, it was frustrating to the people involved in the community. Left with no alternative, naturally the community turned to the Congressional office to try to find some solutions.

I would have to say that I think in this instance the industry has responded, trying to put together an effective plan among a number of companies to get the job done, both with respect to dealing with that specific problem, and I think a lot is going on to deal with the problem generally.

Nevertheless, despite all of the environmental laws, rules and regulations, there seems to be a serious loophole in the law on aboveground storage tanks. We regulate underground storage tanks which leak. We regulate contaminants that seep into streams. But we don't grant any regulatory authority to the EPA to clean up or regulate leaking aboveground storage tanks which directly contaminate the groundwater. Frankly, that doesn't seem to make a lot of sense to me.

I am not one for more Federal regulation than we need. But I think this is a problem that is unresolved that needs attention, and whether this is the exact bill that needs to be enacted or whether the industry is going to be able to demonstrate a response that would lessen the need for regulation, I think remains to be seen. This kind of leakage is a serious threat to human life and a safe environment and something that I think we need to really look at seriously.

At a minimum, regulation should provide for new tank construction standards, testing of existing structures, spill protection and monitoring protocols. In particular, given the serious technical problems associated with groundwater cleanup, we should ensure groundwater protection by requiring both primary and secondary containment that guarantee a high standard of impermeability.

Finally, Congress should take a serious look at how best to address those areas which already have significant contamination. Currently, there are no requirements which are enforceable to clean up the contamination from a leaking aboveground storage tank. It would be irresponsible not to choose to address this question should Congress choose to act on this matter.

Again, I want to thank the chairman for his indulgence of a member of the committee and stress I believe this is an issue which deserves our serious consideration.

I have, Mr. Chairman, some further material which, with unanimous consent, I would like to submit for the record.

Mr. SWIFT. Without objection.

Mr. MCMILLAN. And yield back the balance of my time.

Mr. SWIFT. I thank the gentleman.

Mr. SWIFT. We are very happy now to recognize the author of the legislation. The gentleman has brought this matter strongly to the attention of the committee.

We welcome you, Jim, and I will ask unanimous consent that yours and the prepared text of all of our witnesses this morning be made a part of the record. Without objection, so ordered. And you may proceed as you wish.

STATEMENT OF HON. JAMES P. MORAN, A REPRESENTATIVE IN CONGRESS FROM THE STATE OF VIRGINIA

Mr. MORAN. Thank you very much, Chairman Swift, and it is good to have Mr. Oxley and Mr. McMillan here as well.

I want to start out by—with a personal note of appreciation for you having the hearing and focusing on this issue. I know the agenda is getting very tight at this time of the year. I would also like to say that I join the chorus of people regretting the fact that you are leaving us, Mr. Chairman. It is a personal loss for a lot of Members and it is a great institutional loss.

Mr. SWIFT. It is good to know you are not joining the other chorus which is cheering.

Mr. MORAN. I think the one that regrets you leaving is drowning out any cheering that might take place.

You have got all the major players on this issue here today: Mr. DiBona, the American petroleum Institute; you have got Clark Houghton, Petroleum Marketers Association of America; Lois Epstein, the Environmental Defense Fund; Tony O'Neill, National Fire Protection Association; Jim Tidwell, the International Fire Code Institute; Marshall Mott-Smith with the Florida Department of Environmental Protection.

We thought that might be interesting to the EPA Administrator. So these are all the people that clearly need to be heard from to develop a responsible piece of legislation.

I introduced H.R. 1360 last year in response to a leak that occurred just 20 miles south of here in Fairfax, Virginia, at the Pickett Road tank farm. There are a number of members that actually live in that area. Fortunately, or perhaps unfortunately for the issue, none of them happen to live in the neighborhood that was immediately adjacent to the tank farm.

But that tank farm called the Pickett Road tank farm that is owned by Star Enterprise, an affiliate of Texaco, was first detected in September of 1990, resulted in approximately 200,000 gallons of aviation oil, diesel oil and gasoline penetrating the soil. It seeped into the groundwater, and it spread beneath the nearby residential community.

As Mr. McMillan said happened in his community, the residents suspected something was wrong, and initially they were stonewalled as well. But now many residents have had to be evacuated from their home because the petroleum vapors reached toxic levels. Many have complained of respiratory problems stemming from the leak. Since it has been fully identified, many more residents have had to permanently abandon their homes.

Star Enterprise has already spent about $200 million in compensation to homeowners, $50 million for medical and other damages and $150 million for decreased property values. So, again, this is an expensive thing that would be far less expensive if we could just prevent these kinds of spills in the first place.

It is the largest leak in this area, but it is certainly not the only one. We have smaller amounts of petroleum that are being cleaned up, the Shell, the Exxon and the Crown facilities in Springfield, Virginia. There is an aboveground storage tank just to the south where Shirley Highway meets with the Beltway. That is not so bad, and they have been responsible about it.

But as we find with about half of the aboveground tank farms all over the country, they are leaking. And any tank farm that was constructed before the mid-1970's, there is a very good chance that leaking is going on.

Virginia is the 16th State to have passed aboveground storage tank regulations, and other States are in the process of enacting comprehensive aboveground tank legislation. But there is something of a patchwork that is going on at State initiative.

By the Environmental Protection Agency's own estimates, there were more than 6,000 reported spills from aboveground tanks between 1988-1990. They released about 14 million gallons of oil. Just to put that in perspective, the 1989 spill from the Exxon Valdez into Prince William Sound released 11 million gallons of oil.

So from 1988 to 1990, that same period, there were 14 million gallons released from these leaking storage tanks. And yet it obviously hasn't received anywhere near the kind of attention that the Exxon Valdez spill generated. And that is because the leaks are slow, and they are underground. So until a residential community comes up with some kind of medical problems or you have a fire disaster, whatever, you are not aware of it.

But there are very serious health, safety and environmental risks that are taking place right now. They have the potential of permanently contaminating groundwater, which is the source of drinking water for more than half of the Nation. In many cases, groundwater contamination is leading to serious surface water contamination, and yet we can't regulate groundwater contamination until it contaminates the surface water. That is the problem.

What is astounding is that, while underground storage tanks are highly regulated, it is the aboveground storage tanks which store 100 billion gallons of oil nationwide. They are very loosely regulated by a patchwork of weak regulations. The GAO concluded that the current laws for aboveground storage tanks are grossly inadequate.

Since the legislation was passed to regulate underground tanks, there was a tendency to purchase aboveground tanks—a natural reaction to the regulations of underground. Really what should have happened is that we should have covered both at the same time.

At the most elementary level, current law doesn't even require comprehensive data collection to know how many aboveground storage tanks are leaking. The legislation that we have is, in fact, reactive. It is not proactive in any way.

While the EPA has comprehensive regulatory authority to regulate underground tanks under the Resource Conservation Recovery Act and operate an entire office devoted to underground tanks, aboveground storage tanks are regulated under the emergency response division whose focus is on spills and not leaks. In other words, focuses on the problem after it has happened, rather than any kind of preventive measures.

There is—there are no credible requirements for prevention of spills such as happened in Fairfax and, in fact, around the country. What Federal authority there is to regulate aboveground storage tank facilities is not designed to prevent underground releases, nor is there specific authority to address existing contamination and ensure no off-site migration of these leaks. That is why we introduced H.R. 1360.

The key to the bill is prevention. The legislation requires the EPA Administrator to issue regulations which would set performance standards for new and for rebuilt tanks designed to prevent releases. Regular inspections and spill and overflow prevention would also be required by the legislation to minimize the impact of leaks which do occur. Release detection systems would be required. Tank owners would be required to develop corrective action plans and obtain insurance to ensure that they can assume their financial responsibility in the event of leaks.

Finally, to fill the vacuum in existing data, owners would be required to report information about their tanks and any releases that exceed 42 gallons. Similar to other environmental programs, if a State applies to have primary jurisdiction over tanks, the EPA, after reviewing and approving their program, can give States that authority.

The legislation includes a requirement that tank owners pay a nominal fee to a Federal or the designated State agency based on the size of the tank. States could receive these fees if they had an EPA-approved program.

Now, some are saying that this bill goes too far and that industrywide standards are more than capable of dealing with leaks and spills. In fact, if they were implemented across the board, they would be.

The American Petroleum Institute—I am sure you are going to hear this in a few minutes—maintains more than 400 standards relating to petroleum storage tanks. But they are all voluntary in nature. They are suggestive. It would be prudent for tank owners to follow the API standards, but they are not forced to do so. As a result, there have been major releases in every single State of this Nation, from Anchorage, Alaska to Austin, Texas, to Syracuse, New York, across the country. You name the State and we can point out a leak.

Obviously, many tank owners are choosing not to follow the American Petroleum Institute standards. And whether you have a large tank or a small tank, a marketing tank, a refinery tank, there is a possibility that there will be a leak. It only makes sense that the tank owners prevent those leaks before they occur.

Senator Robb is the chief sponsor of the bill in the Senate. The GAO will be reporting to the Congress on the number and types of aboveground storage tanks across the country. While the EPA is working to ascertain this information, it is important that we immediately find answers to these questions and develop policies to prevent these damaging leaks into the environment.

So, in summary, there is no question that aboveground tank leaks are clearly pervasive across the country, and they are ongoing as we speak.

I am anxious to work with the committee and the EPA and the regulated industries to address the current deficiencies in Federal law and, hopefully, to use H.R. 1360 to fill some of these gaps. And so, in so doing, Congress can address one of the last remaining loopholes in Federal environmental law.

And I very much appreciate the opportunity to testify before you today.

Mr. SWIFT. Thank you very much, Jim. It is a very comprehensive statement.

I want to commend you for the work you have done on this issue, the development of the legislation and particularly for bringing it to the committee's attention so that we were able to hold this hearing and which I hope will lead to further action on this proposal, if not in this Congress, certainly in future congresses.

Recognize the gentleman from Ohio.

Mr. OXLEY. I have no questions.

Mr. SWIFT. Recognize Mr. McMillan.

Mr. MCMILLAN. No questions.

Mr. SWIFT. Thank you, Jim, very, very much. We greatly appreciate your participation.

Mr. MORAN. Thank you, Mr. Chairman.

Mr. SWIFT. Would Mr. Moorhead, the ranking Republican of the committee, care to make a statement at this time?

Mr. MOORHEAD. I do not have an opening statement this morning, but I certainly welcome our witnesses today.

Mr. SWIFT. Thank you very much.

Mr. SWIFT. With that, we welcome Mr. Peter Robertson, Deputy Assistant Administrator of the Office of Solid Waste and Emergency Response of the EPA. Good to see you again.

Mr. ROBERTSON. Thank you, Mr. Chairman.

Mr. SWIFT. Your statement has already been made a part of the record, and you may proceed as you wish.

STATEMENT OF PETER ROBERTSON, DEPUTY ASSISTANT AD-
MINISTRATOR, OFFICE OF SOLID WASTE AND EMERGENCY
RESPONSE, ENVIRONMENTAL PROTECTION AGENCY, AC-
COMPANIED BY DEBBIE DIETRICH, DIRECTOR, EMERGENCY
RESPONSE DIVISION, OFFICE OF EMERGENCY AND REME-
DIAL RESPONSE

Mr. ROBERTSON. Mr. Chairman, I have asked Debbie Dietrich, who is the Director of the Emergency Response Division within the Office of Emergency and Remedial Response at EPA, to join me at the table. I appreciate your welcome.

I am pleased to appear before the subcommittee today to discuss aboveground storage tanks and, in particular, H.R. 1360, Mr. Moran's Safe Aboveground Storage Tank Act of 1993.

EPA has been engaged in regulatory activities dealing with aboveground oil storage facilities for more than 2 decades, Mr. Chairman. Our activities in this area were initiated and are still conducted under section 311 of the Clean Water Act, which author-izes the President to protect surface waters from the releases of oil and hazardous substances.

I should make clear that neither the plain language of this stat-ute as it was amended by the Oil Pollution Act nor the legislative history explicitly authorizes EPA to protect groundwater in cases where there is no discharge or threatened discharge to surface water. Authority to prevent spills by implementing section 311 is divided among the Department of Transportation, the Department of the Interior and the Environmental Protection Agency. EPA's ju-risdiction encompasses inland non-transportation-related facilities.

Pursuant to section 311, EPA requires owners and operators of non-transportation-related facilities to develop and implement so-called spill prevention control and countermeasures, or SPCC, plans. The facilities that must implement these plans include refin-eries, tank farms, fuel oil dealers and many users of petroleum and nonpetroleum products.

Very briefly, the SPCC regulation requires the use of good engi-neering practices in the design, construction, operation and mainte-nance of affected facilities. Not only must every facility subject to the regulation have an SPCC plan but such facilities must also have the procedures, the equipment and the trained personnel needed to implement the plan.

Under the Oil Pollution Act of 1990, which amended section 311 of the Clean Water Act, owners and operators of facilities that could cause substantial harm to the environment are now required to prepare plans for responding to worst-case discharges or the threat of such discharges. The OPA greatly strengthened EPA's au-thority to impose penalties for violations and strengthened our re-sponse authority by making it clear that facility owners and opera-tors have the primary responsibility for cleanup and by creating a $1 billion trust fund that, subject to congressional appropriations, EPA and other Federal agencies can use where necessary to con-duct cleanup activities.

Over the past 4 years, in response to the Agency's own recogni-tion of needed improvements and in accordance with the Oil Pollu-tion Act, EPA has begun to strengthen its regulatory requirements. A rule-making proposal in October 1991 clarified that elements of an SPCC plan are mandatory, and it required, among other things, that owners and operators of affected facilities furnish EPA with information on facility location and size, as well as certain other data.

Based on comments on this proposed rule, EPA initiated a survey to gather data about the regulated universe of facilities. The final rule will reflect the results of our ongoing data collection efforts.

Leaks from oil storage facilities, similar to the one in Congressman's Moran's district, have occurred in many other communities. The Environmental Defense Fund documented many of these incidents in an excellent report issued in February 1993.

In addition, the American Petroleum Institute recently completed a survey of member companies' facilities. Of the facilities that participated in the survey, 81 percent of those that monitor groundwater have confirmed groundwater contamination. Whether this contamination is due largely to historical practices that are no longer followed, as API maintains, or due in part to more recent releases is not clear. In any case, the data do confirm that contamination exists at these facilities.

In 1993, Administrator Browner created an EPA work group to perform a strategic review of aboveground oil storage facilities. In the course of that review, the work group held forums in Philadelphia; Austin, Texas; San Francisco and here in Washington to hear from all stakeholders. Discussions at the forums focused on the nature of the problem and potential solutions. These discussions illustrated that a wide range of opinion exists with respect both to the existence of a problem as well as potential solutions.

I think it is fair to say that while we know a great deal about the occurrence of oil spills into surface waters, we still have a lot to learn about the occurrence of soil and groundwater contamination at aboveground oil storage facilities. In part, the gaps in our knowledge reflect the fact that existing law does not explicitly require that such contamination be reported to the Federal Government. In addition, aboveground oil storage facilities are numerous, diverse and geographically widespread.

EPA estimates that about 500,000 aboveground oil storage facilities may be subject to our SPCC requirements. The EPA facilities' survey, which is aimed at defining and characterizing the universe of EPA regulated aboveground oil storage facilities, is currently underway, as I mentioned. Even when that is completed, however, we believe it will be necessary to collect additional data to define the actual extent of oil contamination of soil and groundwater and the impacts on human health and the environment.

Earlier this year, in a letter from the Administrator to you, Chairman Swift, the EPA described our specific comments on H.R. 1360. Let me briefly reiterate the main points.

First, it is essential that States be fully involved in any effort to deal with aboveground storage facilities and that any legislation calling for delegation to the States provide funding to enable them to play their appropriate role.

Second, several States have already undertaken ambitious efforts to deal with soil and groundwater contamination at aboveground storage facilities. We believe that these States should not be required to make major modifications to their programs in order to conform to any new Federal requirements.

Third, any new legislation dealing with discharges from aboveground storage facilities should build on existing laws and regulations and should specify how its provisions are related to the Clean Water Act and to other Federal laws that pertain to such discharges.

Fourth, such legislation must clearly apply not just to tanks, we believe, but to all the structures, equipment and activities at aboveground storage facilities. That is, the legislation should apply to the facility itself, not just to the tanks, so that all significant sources of releases can be addressed.

Fifth, because we believe there is still much to be learned about the problems that such legislation would address, the legislation should give EPA and States the flexibility to modify their activities as new knowledge emerges.

Sixth, it should give EPA and States latitude to take a risk-based approach to dealing with aboveground storage facilities so that both public and private sector resources can be directed towards the most serious health and environmental threats.

Seventh, because the universe of facilities is large and is diverse, EPA and the States must be able to employ innovative ways of assuring that owners and operators take steps necessary to prevent, detect and clean up discharges.

And, eighth, such legislation must include adequate provisions for penalizing violators.

Mr. Chairman, that completes my oral presentation, and I will be happy, with Ms. Dietrich, to try to answer any questions that you may have.

[The prepared statement of Peter Robertson follows:]

STATEMENT OF PETER ROBINSON, DEPUTY ASSISTANT ADMINISTRATOR, OFFICE OF SOLID WASTE AND EMERGENCY RESPONSE, ENVIRONMENTAL PROTECTION AGENCY

Good morning, Mr. Chairman and members of the subcommittee. I am Peter Robertson, Deputy Assistant Administrator for Solid Waste and Emergency Response at the Environmental Protection Agency (EPA). I am pleased to appear before the subcommittee today to discuss aboveground storage tanks and in particular H.R. 1360, the Safe Aboveground Storage Tank Act of 1993.

I know that the administrator shares the very real concerns that prompted Congressman Moran to introduce H.R. 1360. Sizable releases of gasoline and other petroleum products from a number of aboveground storage facilities—including one in Congressman Moran's district—have demonstrated that such facilities have the potential to cause significant ground water contamination, disrupt the lives of people living near these facilities, and threaten their health.

EPA has been engaged in regulatory activities dealing with aboveground oil storage facilities for more than 2 decades. Our activities in this area were initiated and are still conducted under section 311 of the Clean Water Act which, in general terms, authorizes the President to protect surface waters from releases of oil and hazardous substances. Authority to prevent spills by implementing section 311 is divided among the Department of Transportation, the Department of the Interior, and EPA. EPA's jurisdiction encompasses inland non-transportation-related facilities.

Pursuant to section 311, EPA requires owners and operators of non-transportation-related facilities to develop and implement Spill Prevention, Control, and Countermeasures, or SPCC, plans. Such facilities must have an SPCC plan if they have aboveground oil storage capacity greater than 660 gallons in a single container, total aboveground oil storage capacity greater than 1,320 gallons, or underground oil storage capacity greater than 42 thousand gallons. Such facilities include refineries, tank farms, fuel oil dealers, and many users of petroleum and nonpetroleum products.

Very briefly, the SPCC regulation requires the use of good engineering practices in the design, construction, operation, and maintenance of affected facilities. Not only must every facility subject to the regulation have an SPCC plan, but such facilities also must have the procedures, equipment, and trained personnel needed to implement the plan. The SPCC requirements have helped prevent many millions of gallons of oil from reaching surface water.

When a spill from a regulated facility does reach surface waters or adjoining shorelines the person in charge of the facility is required to notify the National Response Center, which is operated by the Coast Guard. Working with the Coast Guard, EPA On-Scene Coordinators direct or monitor clean-up operations at hundreds of inland spills annually.

Under the Oil Pollution Act of 1990 (OPA), which amended section 311, owners and operators of facilities that could cause substantial harm to the environment are now required to prepare plans for responding to worst-case discharges or the threat of such discharges. OPA greatly strengthened EPA's authority to impose penalties for violations and strengthened our response authority by making it clear that facility owners and operators have the primary responsibility for clean-up and by creating a $1 billion Trust Fund that EPA and other Federal agencies can use where necessary to conduct clean-up activities.

EPA's activities under the Clean Water Act and the Oil Pollution Act have been focused on protection of surface waters. Neither the plain language of these statutes nor their legislative history explicitly authorize EPA to protect ground water in cases where there is no discharge or threatened discharge to surface water.

Over the past 4 years, in response to the Agency's own recognition of needed improvements, and in accordance with OPA, EPA has begun to strengthen its regulatory requirements. A rulemaking proposal in October 1991 clarified that certain elements of an SPCC plan are mandatory and required, among other things that owners and operators of affected facilities furnish EPA with information on facility location and size, as well as certain other data. Based on comments on the proposed rule, EPA initiated a survey to gather data about the regulated universe of facilities. The final rule will reflect the results of these ongoing data collection efforts.

EPA will continue to be vigilant in preventing and cleaning up oil spills that affect surface waters, but it is apparent that we must also pay increasing attention to the threat—and the reality—of oil contamination of soil and ground water. Episodes similar to the one in Congressman Moran's district have occurred in many other communities. The Environmental Defense Fund documented many of these incidents in a report issued in February 1993. In addition, the American Petroleum Institute (API), which recently completed a survey of member companies' facilities, reported just 2 months ago that—of the facilities that participated in the survey— 81 percent of those that monitor ground water have confirmed ground water contamination. Whether this contamination is due largely to historical practices that are no longer allowed, as the API maintains, or due in part to more recent releases, is not clear. In any case, the data suggest that contamination exists at these facilities.

In 1993 the administrator created an EPA Work Group to perform a strategic review of aboveground oil storage facilities. In the course of that review, the Work Group held forums in Philadelphia, Austin, San Francisco, and here in the Washington area to hear from representatives of petroleum producers and marketers, State and local governments, environmental and community groups, and other stakeholders. Discussions at the forums focused on the nature of the problem, if any, and potential solutions. These discussions illustrated that a wide range of opinion exists with respect to both the existence of a problem as well as a potential solution. A summary of the highlights of the proceedings of these forums has been released, and I would be happy to provide a copy for the record of this hearing, if you wish to include it.

I think it is fair to say that while we know a great deal about the occurrence of oil spills into surface waters, we still have a lot to learn about the occurrence of soil and ground water contamination at aboveground oil storage facilities. In part, the gaps in our knowledge reflect the fact that existing law does not explicitly require that such contamination be reported to the Federal Government. In addition, aboveground oil storage facilities are numerous, diverse, and geographically widespread. EPA estimates that about 500,000 aboveground oil storage facilities may be subject to the SPCC requirements. As mentioned above, the EPA facilities survey which is aimed at defining and characterizing the universe of EPA regulated aboveground oil storage facilities is currently underway. Even when that is completed, however, it will be necessary to collect additional data to define the actual extent of oil contamination of soil and ground water and the impacts on human health and the environment.

EPA shares Congressman Moran's interest in defining the nature of the problem as well as the role the Federal Government should play in preventing and remediating ground water contamination resulting from the operation of aboveground oil storage facilities. Earlier this year, in a letter from the administrator to Chairman Swift, we described our specific comments on H.R. 1360. I will briefly reiterate the main points:

First, it is essential that States be fully involved in any effort to deal with aboveground storage facilities and that any legislation calling for delegation to States provide funding to enable them to play their appropriate role.

Second, several States have already undertaken ambitious effort to deal with soil and ground water contamination at aboveground storage facilities; States should not be required to make major modifications to these programs in order to conform to Federal requirements;

Third, any new legislation dealing with discharges from aboveground storage facilities should build on existing laws and regulations and should specify how its provisions are related to the Clean Water Act, as amended by the Oil Pollution Act, and to other Federal laws that pertain to such discharges;

Fourth, such legislation must clearly apply not just to tanks but to all structures, equipment, and activities at aboveground storage facilities so that all significant sources of releases can be addressed;

Fifth, because there is still much to be learned about the problem that such legislation would address, it should give EPA and States flexibility to modify our activities as new knowledge emerges;

Sixth, it should give EPA and States latitude to take a risk-based approach to dealing with aboveground storage facilities, so that both public and private sector resources can be directed toward the most serious health and environmental threats;

Seventh, because the universe of facilities is large and diverse, EPA and States must be able to employ innovative ways of assuring that owners and operators take the steps necessary to prevent, detect, and clean up discharges; and

Eighth, such legislation must include provisions for penalizing violators.

In summary, legislation addressing contamination from aboveground storage tanks should provide opportunities to improve our knowledge and modify our approach as our knowledge increases; build on our two decades of experience in dealing with aboveground oil storage facilities; offer flexibility to tackle the problem on the basis of an objective assessment of health and environmental risks; and encourage, and provide support for State involvement. EPA is prepared to work with this subcommittee and other stakeholders to fashion an appropriate solution to the problems of soil and ground water contamination associated with aboveground storage facilities. I will be happy at this point to answer any questions the subcommittee may have. Thank you.

Mr. SWIFT. Thank you very much.

I just want to note in toward the very end of your statement that you mentioned that there needed to be some inclusion of risk assessment involved in all of this, and I approve of that. I think some who have decided that risk assessment is for, whatever reasons, automatically a bad thing need to reassess. I think risk assessment has got to become a major tool that we use.

The caution on the other side is that it not become an article of religious faith, as well. And I—in listening to some of the proponents of risk assessment, I kind of hear that developing, too, that it is a grand magic wand that will solve all problems, and I don't believe that.

If we can wrestle this debate away from the extremes and concentrate on risk assessment that makes sense and is used as a major tool in the toolbox that we use to try to fix environmental issues, it can be very, very good. And I am happy to note that EPA believes that risk assessment has a role in these issues.

Let us talk about your authority under the Oil Pollution Act. That is tied to contamination of surface water so that a release of contaminants that only goes into groundwater falls into a regulatory gap. That is, obviously, a concern for you. Do you find this legislation effectively deals with that or would you rather approach it in some other way?

Mr. ROBERTSON. Certainly, Mr. Chairman, Mr. Moran's legislation would close that gap. I believe it is the single biggest gap in our existing authority that we recognize and that prevents us from dealing with discharges to groundwater and to the soil.

Let me also point out that virtually every State has some sort of authority to deal with groundwater contamination, and a number of States are dealing with these sorts of problems under their own authorities now.

Mr. SWIFT. Do you have any authority to require notification of underground leaks from aboveground tanks?

Mr. ROBERTSON. We do not, Mr. Chairman, have that authority, either.

Mr. SWIFT. What I think we are going to hear later today is that there is nothing to worry about because it is all being taken care of one way or another, and I don't think that has got any credibility. Likewise, the concern that we are about to write some great, grandiose thing that will just, you know, devour everybody is not an illegitimate concern.

Do you think, whether it is the Moran proposal or something else, that there is a way to come up with a unified regulatory scheme that may even be simpler in that those affected by it will have a—a one-stop-shop kind of situation where they would need to go? Or because of the need to involve States and what have you, do you just end up with a different kind of complex regulatory scheme?

Mr. ROBERTSON. Mr. Chairman, the working group that Administrator Browner formed in April of last year has been looking at this issue. Their report is—frankly, should be released in about the next week or so. It is being reviewed by senior management even as we speak. I am prepared to talk a bit about the recommendations of that report, and I think it addresses your questions.

Assuming that Administrator Browner approves their report, the working group will make four general suggestions for moving forward on this problem.

First, we believe we do have to continue data collection to better analyze and characterize the universe of these facilities, their operations and the soil and groundwater contamination that are coincident to those operations.

We also believe that we need to build on our existing regulatory structure under section 311, particularly by completing the SPCC rule amendments which were proposed in 1991 and which have been delayed, Mr. Chairman, while we have responded to the other mandates of the Oil Pollution Act of 1990 and other regulatory requirements. And also, frankly, of late we weren't rushing to get these out because we wanted to give the aboveground storage tank work group an opportunity to make an independent assessment. But we do believe we can strengthen our regulations and that the October 1991 proposal will do so.

Number three, we are willing to continue to work with this committee and other committees in Congress in dealing with any legislation that might be proposed, whether it be an approach like Mr. Moran's that provides a new regulatory regime or whether it be amending the Clean Water Act to broaden our existing authority so that we can address some of these problems.

Finally, we want to investigate the feasibility of developing voluntary programs in cooperation with the industry to both clean up current contamination and prevent future spills. In that light, Mr. Chairman, I know that you are aware of the Common Sense Initiative which is one of the highest priorities of Administrator Browner right now. Under the Common Sense Initiative, we are looking at six different industrial sectors to determine how we can do things cleaner, cheaper and smarter.

The petroleum refining industry is one of the six sectors that we are working with right now. I think this Common Sense Initiative may provide an outstanding opportunity for us to work with the petroleum industry to see whether, in addition to the other things that I have described, we might be able to augment both cleanup and prevention by working voluntarily with the industry.

So we are prepared to proceed on all four of these working group suggestions, Mr. Chairman, to try to deal comprehensively with a problem that we believe is a serious one.

Mr. SWIFT. I think one of the things that wears you down when you sit on this side of the dais is that everything seems to start with those who want to do everything yesterday and those who want to not do anything ever.

It seems to me that under Carol Browner EPA has been trying more and more to include affected parties in the development of solutions which I think has—is laudable and is patterned after at least some of the concepts, I think, that are used by some governments in Europe to try and bridge this adversarial relationship that too often exists.

It seems to me that you have something you can offer affected people. It seems to me that business, if they can get certitude, if they can get rapid turnaround, standards do not become as difficult to achieve if industry knows that it can get its decisions made quickly and effectively and consistently. That we really need to remodel our regulatory structure so that, in effect, for those of us who are concerned about the public health and safety can establish standards we need and then reward business, if you will, by saying, you know, we are going to make this as easy to comply with as we possibly can.

I know that is a concern of Carol Browner's, and I think that the effort that is under way on this issue may lead that way. If business gets a little skeptical about that at the outset, I don't blame them. I am from the Federal Government; I am here to help you kind of thing.

But unless they think the environmental movement and all those concerns are going to disappear any time in their lifetime, which would be naive beyond the wildest dreams, it seems to me that it is in industry's interest to help find a better model than this terrible adversarial process that we use constantly here which leads to winner-take-all kinds of decisions, which are usually not the best decisions and so forth and so on.

With Miss Browner in the lead, I think we are searching for some models and I hope there will be some response, positive response, from industry to that, even though they may want to keep their guard up for a little while, see how it all works out.

One last question which really grows out of all of this. When we talk about cost-benefit, which is a very big thing, sometimes we never talk about the benefits from certain regulatory structures. And it seems to me that here, while there is going to be some costs, that you have also got depreciation of property value that comes from leaks. You have got health and medical expenses that can be very costly. Has EPA got any study of what the benefits would be to having a better regulatory structure for dealing with above-ground storage tanks?

Mr. ROBERTSON. Well, Mr. Chairman, certainly one study that will touch upon that issue is the liner study that was required as part of the Oil Pollution Act of 1990. We have had the liner study under way for some time, and I believe we are relatively close to releasing that as well. It has been deferred for some of the same reasons that I mentioned about the SPCC regulatory changes.

That study, while it is not ready for release yet so I am not prepared to talk about specific numbers at this point, does address costs and benefits of various options like protective liners of double bottoms for tanks.

Mr. SWIFT. Thank you very much.

The Chair will recognize the gentleman from Ohio.

Mr. OXLEY. Thank you, Mr. Chairman.

Mr. Robertson, has the EPA had a chance to implement or to decide what kind of resources would be necessary to implement the legislation that Mr. Moran has introduced?

Mr. ROBERTSON. To the best of my knowledge, Mr. Oxley, we have not yet put a figure on what it would cost to implement. In the letter that I mentioned to Chairman Swift, we expressed some of our concerns about the legislation. We obviously hope those concerns could be accommodated in any ongoing legislative debate. But I can't tell you what we think it would cost to implement the——

Mr. OXLEY. Was one of the concerns the cost factor?

Mr. ROBERTSON. That is one of the concerns. It is a concern not only for EPA—we are always trying to do more with less, Mr. Oxley—but it is also a concern for the States.

One of our concerns about H.R. 1360 which was specifically identified in Administrator Browner's letter to Chairman Swift was that this provides a new requirement for States but doesn't provide funding to help them implement it. It doesn't have anything similar to the trust funds provided in the underground storage tank program and Oil Pollution Act that are used to both clean up sites when you can't find potentially responsible parties as well as to help fund the State implementation of the programs. So certainly cost concerns both from EPA's perspective as well as the States' perspective exists.

Mr. OXLEY. From the States' perspective, that is a clear example of an unfunded mandate as it exists now, is that correct?

Mr. ROBERTSON. Mr. Moran's legislation does provide for an authorization of appropriations of such funds as may be necessary in order to deal with this new regulatory regime. Obviously, it doesn't make appropriations in the bill.

Mr. OXLEY. Does Mr. Moran's bill consolidate existing Federal authorities or does the bill overlap the authority of the current SPCC requirements?

Mr. ROBERTSON. Well, another of the concerns that Administrator Browner identified, Mr. Oxley, is that H.R. 1360 doesn't entirely make clear how this new regulatory regime would fit in with the Clean Water Act and other existing authorities. So that is also a continuing concern with the legislation.

Mr. OXLEY. How can we be assured that existing State programs are not disrupted under this legislation?

Mr. ROBERTSON. Again, another concern that Administrator Browner has identified and one of our suggestions for the legislation would be that it makes clear that the States that have already started vigorous aboveground storage tank programs don't have to revise those programs in a major way in order to comply with new Federal mandates.

Mr. OXLEY. So it would appear that the fact that this bill is not going anywhere is probably good news for not only the industry but the States and EPA as well.

Mr. ROBERTSON. As I mentioned, we will have four general recommendations from the aboveground storage tank work group. One of those recommendations is additional legislation. While we have identified concerns about Mr. Moran's legislation, we do think that additional legislative authority is going to be needed ultimately to help us adequately deal with this problem. And I am not here to say flatly that Mr. Moran's version is not the answer or that it is the answer. We have concerns with the legislation, and we would like to see those concerns addressed.

Mr. OXLEY. It is clearly not the answer today because it is not going to pass. And so the question is what we do in the next intervening months or what EPA does to help bring some, as you actually said, common sense to this kind of a solution. And I applaud the Administrator for her vision in dealing with these kinds of issues and for setting up an agenda that is indeed a common sense agenda, and certainly I am encouraged by that.

Thank you, Mr. Chairman.

Mr. SWIFT. The gentleman from North Carolina.

Mr. MCMILLAN. Thank you, Mr. Chairman.

I compliment you on your statement about handling situations such as this. I think your approach to this is going to be missed. Hopefully, the succeeding leadership of the committee will be equally balanced in their approach to dealing with this—and the administration, as well.

It seems like that so often we do operate around here on lines in the sand. It has come up time and time again. And yet when we get adversaries together and try to come up with a common solution, particularly on a pragmatic problem, in a closed room, we get pretty far along in being able to do that.

And then we get to the point where it may end up being voted on and the groups subdivide again because they think they might come out better if they hold out a little bit longer and so forth, and we don't get things done.

Mr. SWIFT. You are not talking about Superfund, are you, by any chance?

Mr. McMILLAN. Well, let's just put it down as all of the above over my 10 years of experience.

And I think that the approach that you are developing sounds certainly sensible if it is really implemented in a way that is convincing to all sides. And I think that is going to be the key to it.

You know, I think it is hard to reinvent government, even if we are serious about it, and I think we are serious about it. However, I am not sure we have got the right answers. But—I mean, this is a perfect example of it.

And if one of those prongs which would include industry cooperation in terms of—because they are the ones that ultimately are going to have to do it. They are probably the ones that are going to have to, in this environment, pass the cost on to the consumer because we are probably not going to pay for it out of tax funds. If we do, it will only be so on the margins.

The cost of the cleanup, the cost of adequate protection in the future is going to become a cost of the product to the consumer. And they have got every interest in solving the problem.

I agree with the chairman, the concerns about storage tanks aren't going to go away. And not just because the Environmental Defense Fund is leading an advocacy position on a national scale. It gets down to the grassroots, because somebody has a pollution problem they are concerned about. When neighborhoods are affected, it becomes something very different.

I think we can find solutions to these things and get them done. Perhaps, in many cases, they have already been thought through.

To what extent, as part of your plan, does EPA remain in full consultation with the industry to really uncover examples of success in dealing with either new construction or with preexisting problems that have to be rectified or should be?

Mr. ROBERTSON. We are just beginning this common sense initiative, Mr. McMillan. In response to the concerns that you have just discussed, similar to Chairman Swift's, I attended the initial meetings for each of the six industry sectors that will participate in the Common Sense Initiative. The petroleum refining industry expressed the greatest degree of initial skepticism about whether this program could work. Frankly, I don't blame them for that at all.

Since they committed to Administrator Browner that they were willing to be one of the industries involved in the Common Sense Initiative, they have worked with the utmost vigor and utmost good faith with the Agency.

We are literally just getting the Common Sense Initiative off the ground. We are looking to provide some early successes so that we can show good faith to the industry and so that the industry has something to point to as signals of progress.

We are very excited about the Common Sense Initiative. The API has been nothing but cooperative since we started. We are just beginning to move forward, but I certainly applaud them for their work so far with us, including individual companies like Phillips Petroleum who has been the leader among the industry in working with us thus far.

Everybody in the Agency is very excited about the opportunities that the Common Sense Initiative presents to revise the way the Agency deals with the petroleum refining industry. We are particularly excited in OSWER because we are the co-lead with region six for the petroleum refining sector. I think it holds opportunities for doing great things.

Mr. McMILLAN. Are there things legislatively that Congress can do to reinforce that? So often we hear you don't have jurisdiction to do that and so forth. And these kind of things forestall what I would call creative managerial approaches to solving the problem, if you will, that should preexist any legislation. Legislation and regulatory action that presumes that everybody is going to be a violator should be a last resort.

What do we need to do that enables you, along with the private sector, to take a creative approach to dealing with the problem before it has to be regulated?

Mr. ROBERTSON. I hope you will still be willing to ask that same question some time in the future, Mr. McMillan. We don't have any specifics to recommend in terms of legislative changes right now. We are still forming the stakeholders group that will discuss the petroleum refining section. We haven't identified all the States and environmental groups as well as industry and EPA stakeholders who are going to participate. But nothing is going to be off the table in these negotiations. And we do anticipate the possibility of coming to Congress and saying that we have identified what we think are some very creative ways to do things cleaner, cheaper, and smarter but we need congressional authority in this area or that area before we can proceed.

Mr. McMILLAN. Legislative recommendations would tend to deal with that rather than an adversarial regulatory regime? It would be sort of enabling?

Mr. ROBERTSON. My sense, Mr. McMillan, is that if we come to you looking for something that is adversarial in nature, we will have failed in the Common Sense Initiative which is supposed to be a cooperative venture.

Mr. McMILLAN. Could some of this take on aspects of empowering the agency to enter some sort of agreement or compact with the industry. We have got precedence for doing this kind of thing. We do it in securities regulations which works extremely well. And I wonder if that could be a part?

Mr. ROBERTSON. I think it absolutely could be, Mr. McMillan. We don't have any specifics yet because this is so new. But we are looking at exactly those sorts of things. We are trying to be creative and trying not to be hindered by the way the agency currently does its business or has done it in the past. And we are looking to the industry to tell us what they need and how we could do better.

We are looking to the environmental groups to help us fashion ways to do things better as well. Sometimes people focus on only one of our three goals: Cleaner, cheaper, and smarter. We definitely want to do all three of them. We want to do things cleaner as well as cheaper. We are looking to all the stakeholders to help us identify ways to improve the way we do business and improve the environment at less cost, transactional cost and otherwise.

Mr. McMILLAN. Well, that sounds very good, and I look forward to seeing what you produce.

Thank you, Mr. Chairman.

Mr. SWIFT. Thank you. I might note something else I think Congress can do. If we want agencies to do things with more common sense, and to show a little initiative, we have got to show a little legislative restraint. We don't do the congressional bully boy bit when things are not precisely the way any one of us would envision. I think we tend to overwrite legislation so they can't do anything dumb. And we found that legislation keeps them from doing anything intelligent, too.

And so if we are unwilling to provide a little flexibility for the agencies, then we can't sit up here and complain because the agencies don't react in as creative ways as they might. I thank you very much. I appreciate your testimony.

Our last panel includes Mr. Charles DiBona, American Petroleum Institute; Lois Epstein with the Pollution Prevention Alliance of the EDF; Mr. Clark Houghton, with the Petroleum Marketers Association; Mr. Marshall T. Mott-Smith with the Florida Department of Environmental Protection; Anthony R. O'Neill with the National Fire Protection Association; and Mr. J.L. Tidwell, chairman of the Uniform Fire Code Committee of the International Fire Code Institute.

Your full statements, any attachments thereto, will be included in the record. You may proceed as you want, and I think we will take people in the order I introduced them, which means we will begin with Charles DiBona, president of the American Petroleum Institute.

STATEMENTS OF CHARLES J. DiBONA, PRESIDENT, AMERICAN PETROLEUM INSTITUTE; LOIS N. EPSTEIN, ENGINEER, POLLUTION PREVENTION ALLIANCE, ENVIRONMENTAL DEFENSE FUND; CLARK HOUGHTON, STATE EXECUTIVE, PETROLEUM MARKETERS ASSOCIATION OF AMERICA; MARSHALL T. MOTT-SMITH, ADMINISTRATOR, STORAGE TANK REGULATION SECTION, FLORIDA DEPARTMENT OF ENVIRONMENTAL PROTECTION; ANTHONY R. O'NEILL, VICE PRESIDENT, GOVERNMENT AFFAIRS, NATIONAL FIRE PROTECTION ASSOCIATION; AND J.L. TIDWELL, CHAIRMAN, UNIFORM FIRE CODE COMMITTEE, INTERNATIONAL FIRE CODE INSTITUTE, ACCOMPANIED BY WAYNE SENTER, FIRE MARSHAL, CITY OF AUBURN FIRE DEPARTMENT

Mr. DiBONA. Thank you, Mr. Chairman. I am pleased to present our members' views on H.R. 1360. API represents more than 300 companies involved in all aspects of the oil and natural gas industry. API has a 75-year history of standards setting for the petroleum industry and our standards have been accepted around the world.

Our intent in this program has been to identify problems that need attention and correct them so that government action would not be required. We have tried to do this in the case of groundwater contamination at terminals and other installations using aboveground tankage. Consequently, API is updating individual existing standards affecting ground storage facilities—aboveground storage facilities and have developed a new comprehensive standard for addressing these concerns.

API members voluntarily participated in a survey to better understand the extent of groundwater contamination, the historic causes of the problem, and improvements of recent years. To address the problems of leaks or spills from aboveground storage tanks, industry representatives serving on API committees developed a comprehensive standard addressing the design, construction, maintenance, and inspection of all petroleum terminals and tank facilities associated with refining, marketing, and transportation activities.

Standard 2610, which became available for use this year, updated and assembled all terminal and tank-related standards which used to be separately developed into one document, thereby providing industry and government with a single source for comprehensive guidance. Copies of API Standard 2610, which is this document, have been made available to the subcommittee and I ask that it be made part of the record.

The information in the standard ranges from pollution prevention and safe operating practices at facilities to removal and decommission of tanks. Besides providing a comprehensive, holistic approach to safe and environmentally sound management of tank and terminal facilities, Standard 2610 is written to be consistent with applicable local State and Federal regulation.

In addition to developing this standard, API conducted at EPA's request a groundwater survey of the industry's refining marketing and transportation activities or sectors. The survey was designed to determine the prevalence of groundwater contamination, rank the sources of contamination, document facility improvements, and determine the status of remedial activities. The evaluation of the source of contamination was important to determine the effectiveness of existing practices and standards to identify areas for improvement.

The results indicate that some level of groundwater contamination exists at a number of the facilities surveyed, primarily because of past operating practices.

In virtually all cases where contamination was identified, it is being treated or remediated. Government agencies are overseeing this remediation in all but a few cases. The industry's current environmental performance has improved significantly because of upgraded equipment and revised operating standards.

The data also show that during the past 5 years groundwater contamination has been caused by a variety of sources, although pressurized buried piping has been the most significant source of contamination. In response to this finding and in keeping with its role of standard setting for the industry, API modified the new terminal and tank standard to address piping integrity. Adherence to the standard now requires that operators ensure the integrity of buried piping on a regular basis.

Overall, the survey indicates that continued emphasis on existing and newly adopted industry operating procedures and standards provides the greatest opportunity for environmental improvement. And I am not here trying to say that we have solved all the problems, Mr. Chairman. We know we have problems, but we think these are steps in the right direction. I just want to make that point clear.

Existing law gives EPA and the States authority to ensure that this happens. And I want to assure you that this industry intends to act responsibly to prevent and respond to releases and to work with EPA in the regulatory arena. Consequently, we believe that H.R. 1360 is unnecessary. Moreover, it takes what we believe is an overly prescriptive approach to regulating aboveground storage facilities with inadequate recognition of the diversity that exists among company facilities.

And much of the points made by EPA in the previous testimony are things that we would agree with about the specifics of that; the need for perhaps some additional authorities, but those are not the ones contained in this particular piece of legislation. And we would be, you know, happy to look at and review anything that they may have ongoing, but the basic points they made about trying to let States carry this out and to rely as much as they can on some of these other authorities, and to use our new standard as a basis for some of that additional regulation, as we understand the State of Florida is doing, makes a lot of sense.

Many of API's standards have been incorporated or referenced in laws in many areas. And that provides a kind of flexible way of ensuring that the newest developments are then incorporated in revisions of the standards. And it is a positive, workable way of handling many of these complex problems.

And what we are interested in is making something that works, just as I think everyone in this room is. And we want to work with them.

So we believe that the steps that we have taken and that we will take, and the existing EPA and State authority should be given a chance to work. And we believe that this approach can succeed at the lowest cost to the consuming public while ensuring a high level of environmental performance and improvement in what historically has been a mixed performance.

[The prepared statement of Charles J. DiBona follows:]

22

**Testimony of the American Petroleum Institute
for the Subcommittee on Transportation and Hazardous Materials
of the Committee on Energy and Commerce
U.S. House of Representatives
on Aboveground Storage Tanks
September 14, 1994**

The American Petroleum Institute (API) appreciates the opportunity to submit testimony on aboveground storage tanks. API represents approximately 300 member companies involved in the exploration, production, refining, transportation, and marketing of petroleum and petroleum products. At every stage, crude oil and its many products are stored in aboveground storage tanks. API members are seriously committed to the proper operation and maintenance of these tanks and could be greatly affected should Congress elect to change the laws that govern terminal and tank operations.

API members believe that no additional federal authority is needed to adequately regulate aboveground storage tanks and terminals. The current federal regulatory program, implemented under the Clean Water Act, addresses both prevention and cleanup of releases from aboveground storage facilities. Additionally, API members have been involved in activities in recent years to improve aboveground tank facility operations and achieve a greater degree of environmental protection. This testimony addresses the results of the actions taken by API members and summarizes our views on H.R. 1360, the "Safe Aboveground Storage Tank Act of 1993."

API Standards for Aboveground Storage Tanks

For most of this century, API has played a leadership role in the development of standards for the petroleum industry in the U.S. and around the world. Today, API administers an extensive program of technical standards that cover all phases of the industry's operations, including the storage of crude oil and refined products. These standards are designed to maintain safe operations and prevent releases and are regularly reviewed and revised by industry experts.

The primary aboveground tank standards, listed below, specify design, construction, operation and inspection requirements at petroleum storage facilities in the refining, marketing, and transportation sectors. The standards reflect improvements in technology and operating practices that resulted from increased environmental awareness. The list also reflects the breadth of operational features addressed, and the publication dates reflect the ongoing revisions to these documents. Recent activity has included:

- Publication of *API Recommended Practice 2350, Overfill Protection for Petroleum Storage Tanks*, 1st edition, March 1987, which provides guidance on the development of an overfill prevention program. The second edition of this recommended practice is under development.

- Publication of *API Standard 651, Cathodic Protection of Aboveground Storage Tanks*, 1st edition, April 1991, which gives details of cathodic protection systems for corrosion control.

- Publication of *API Recommended Practice 652, Lining of Aboveground Petroleum Storage Tank Bottoms*, 1st edition, April 1991, which gives details of interior bottom lining systems to control corrosion and other causes of leaks.

- Publication of *API Standard 653, Tank Inspection, Repair, Alteration and Reconstruction*, 1st edition, January 1991. Among other requirements, this standard specifies an internal inspection frequency, based on a calculated corrosion rate. Currently, Standard 653 is referenced in regulation by six states.[1]

- Establishment in 1991 of the Aboveground Storage Tank Inspector Certification Program (ASTICP), which insures qualified inspectors.

- Publication of *API Standard 650, Welded Steel Tanks for Oil Storage*, 9th edition, May 1993, which covers material design, fabrication, erection, and testing of tanks and gives guidance for the installation of release prevention barriers under ASTs.

- Publication of *API Standard 2015, Safe Entry and Cleaning of Petroleum Storage Tanks: Planning and Managing Tank Entry from Decommissioning Through Recommissioning*, 5th edition, May 1994, which provides procedures for taking a tank out of service and safely entering it for cleaning, inspection, and other purposes.

- Publication of *API Standard 2610, Design, Construction, Operation, Maintenance and Inspection of Terminal and Tank Facilities*, 1st edition, July 1994, which provides a comprehensive guide to the best industry practices for terminal design, construction, inspection, maintenance, repair, and environmental protection. (Development of this standard is discussed below.)

In addition to maintaining industry standards, API established a program for improved environmental performance in 1990. Known as Strategies for Today's Environmental Partnership, or STEP, the program established seven sets of management practices including a set on pollution prevention. API members support and participate in this program. [Note: Additional information on the STEP program is attached.]

Evaluation of Aboveground Storage Tank Operations

In 1992, API formed a Terminal and Tank Steering Group, to evaluate the industry's and API's approach to aboveground storage tank facility operation. The group decided on two courses of action. The first involved a reassessment of the adequacy of API standards for terminal and tank operations. The second involved a survey of API members to evaluate the current status of facility operations and determine the frequency of groundwater contamination.

- **Development of a Comprehensive Standard for Terminals and Tanks**

Under the Steering Group's direction, approximately 60 API standards and recommended practices were reviewed. This resulted in the revision of several existing standards and development of new ones. Most importantly, API members developed a comprehensive new standard (Standard 2610), which addresses the design, construction, operation, maintenance, and inspection of all petroleum terminal and tank facilities associated with marketing, refining, and transportation activities. The standard was designed to fill a need to combine terminal- and tank-related standards and good operating practices into one document, which industry and government alike can turn to for guidance.

The new standard provides a comprehensive, holistic approach to safe, environmentally sound management of tank and terminal facilities and is intended to be consistent with applicable local, state, or federal regulations. The standard covers such subjects as site selection and spacing, pollution prevention, waste management, safe operating practices, fire protection, dikes and berms, mechanical systems, product transfer, corrosion protection, utilities, as well as removal and decommissioning of tanks. Overall, it stresses environmental protection.

In addition to receiving unanimous approval within API, Standard 2610 has been approved by the American National Standards Institute, making it the recognized American National Standard for terminal and tank operations. At least one state, Florida, is in the process of incorporating the standard into its regulations.

API members believe that industry standards, such as the comprehensive terminal and tank standard, as well as continuing industry action to improve design and operating practices mitigate the need for further legislation. The industry wants to identify problems and correct them and has strong economic incentive to do so.

- **API Survey of Aboveground Storage Tank Facilities**

In an effort to further define operational problems at facilities and assess recent improvements, API conducted a survey of its members' aboveground storage tank facilities. At EPA's request, API members in the refining, marketing, and transportation sectors voluntarily participated in a survey of these facilities.[2] The purposes of the survey were to assess recent facility improvements, determine the frequency of groundwater contamination,[3] rank the relative sources of groundwater contamination, and determine the status of remedial activities at API member company facilities. The evaluation of the sources of contamination was particularly important to determine the effectiveness of practices and standards used at facilities and to identify areas for additional improvement.

The survey was sent to a random sample of 350 API member company facilities.[4] Responses were received from 299, or 85 percent, of the sampled facilities. To achieve a high rate of participation, names and locations of responding facilities were kept confidential by the independent survey research firm that conducted the sampling and tabulated the data.

The data indicate that the contribution to groundwater contamination from almost all sources has decreased. Respondents in all three industry sectors reported significant reductions in releases within the past five years because of improved equipment and operating practices. Improvements include:

- Application of upgraded or newly developed API operational standards;
- Implementation of more rigorous inspection programs (as specified by API standards);
- Addition and enhancement of overfill protection systems;
- Installation of tank water-bottom collection systems; and
- Increased utilization of cathodic protection systems for buried piping.

The survey results for each industry sector are reported below:

•Refining Sector Results: An estimated 98 percent of refineries use subsurface monitoring techniques to determine hydrocarbon contamination. Some level of groundwater contamination was reported at 85 percent of refineries.* This contamination is due in large part to past operating practices which have been modified or eliminated within recent years. All of the refineries that reported contamination also reported that remedial activities are taking place at the facility. In virtually all cases, these remedial activities occur under the oversight of a government agency.

•Marketing Sector Results: An estimated 80 percent of marketing terminals use subsurface monitoring techniques to determine hydrocarbon contamination. Some level of groundwater contamination was reported at 68 percent of marketing terminals.* As in the refining sector, this contamination is due in large part to past operating practices which have been modified or eliminated within recent years. Remedial activities are taking place at an estimated 95% of marketing terminals that have groundwater contamination.[5] In virtually all cases, these remedial activities occur under the oversight of a government agency.

•Transportation Sector Results: An estimated 18 percent of transportation terminals use subsurface monitoring techniques to determine hydrocarbon contamination. Some level of groundwater contamination was reported at 10 percent of transportation terminals.* As in the other two sectors, this contamination is due in large part to past operating practices which have been modified or eliminated within recent years. Remedial activities are taking place at an estimated 90 percent of transportation terminals that have groundwater contamination. In virtually all cases, these remedial activities occur under the oversight of a government agency.

[*Note: The existence of groundwater contamination does not necessarily mean that drinking water has been affected.]

The survey data indicate that groundwater contamination exists; however, the data also show that contamination at aboveground storage facilities can be primarily attributed to past operating practices. Upgraded equipment, operating standards, and the management practices included in the STEP program have improved the industry's environmental performance. In virtually all cases, where contamination has been identified, it is, or has been, treated or remediated. Such remediation is taking place, with few exceptions, under the oversight of a government agency.[6]

According to the survey results, groundwater contamination appears to have been caused by a variety of sources during the past five-year time period. The data indicate, however, that pressurized buried piping has been the most significant source of contamination at facilities in all three sectors over the past five years. In response to this survey finding, API modified its new terminal and tank standard (Standard 2610) to address piping integrity. Adherence to the standard now requires that operators assure the integrity of buried piping on a regular basis.

The data also indicate that releases from aboveground storage tank bottoms are not a primary source of contamination. Survey respondents indicated that less than 3.6 percent of tanks (in all age categories) had confirmed bottom failures over the past five years. Furthermore, the importance of tank bottoms as a source of contamination has significantly declined over the past five years in all industry sectors, largely because of improved inspection procedures. Mandates for the universal installation of liners underneath aboveground storage tanks and in secondary containment areas would provide no significant additional protection at these facilities and would be unnecessarily costly.

It is apparent from API's survey of aboveground storage tank facilities that some level of contamination exists at many petroleum industry facilities primarily because of past operating practices--practices that have been modified or eliminated in recent years. The potential environmental impact of these practices was not well understood in the past. However, the petroleum industry, like other industries and government, has learned from experience. Changes have been made in operating practices, standards updated, and new equipment installed to reduce future risk of groundwater contamination.

The survey results indicate that continued industry focus and emphasis on existing and newly adopted industry operating procedures and standards has provided steady improvement in facility operations. Moreover, the survey results demonstrate the petroleum industry's commitment to making continued improvements in operating practices as well as the industry's commitment to working closely with state and local governments.

API Position on Additional Federal Legislation

The Spill Prevention, Control and Countermeasures (SPCC) program, implemented under authority of the Clean Water Act, gives EPA authority to issue regulations governing the cleanup of spills and establishes methods for preventing discharges of oil and hazardous substances. EPA has proposed revisions to the SPCC regulatory program that, once finalized and implemented, will result in more stringent standards. Additionally, under authority of the Oil Pollution Act of 1990, the Agency is currently conducting a study to determine whether liners should be used to prevent releases from onshore petroleum storage facilities. The recommendations of this study are to be implemented within six months of its submission to Congress. API believes that the proposed revisions to the SPCC program should be implemented and the liner study finalized before the need for additional legislative authority can be evaluated fairly.

Current laws require industry to prevent and respond to releases from aboveground storage facilities, and authorize EPA to enforce these requirements. API members believe that they have demonstrated their willingness both to prevent and respond to releases and to work with EPA to periodically reassess the adequacy of existing regulations. A major reason that API undertook its recent survey was to supply data for EPA's liner study.

In light of existing federal laws and ongoing EPA regulatory activities, API members believe that H.R. 1360 is unnecessary. Moreover, as currently written, the bill takes a rigid and overly prescriptive approach to regulating aboveground storage tanks.[7] History demonstrates that performance standards are preferable to engineering dictates, which can be uneconomic, inappropriate, and curb technological development. The bill sets arbitrary mandates that ignore the practical impact such requirements would have on facility operations. For example:

- Requiring existing tanks to meet new tank standards within ten years is unrealistic in light of the size and diversity of the tank universe.

- The frequency of required tank inspections is excessive and unnecessary. The frequency of inspections should be based--as it is in API standard 653--on system engineering considerations such as the site-specific corrosion rate.

- The bill mandates impervious, secondary containment. Such a mandate is clearly premature as EPA's study on the use of liners and other means of secondary containment has not been completed and unwarranted in light of the data contained in API's aboveground storage tank facility survey. The data show that tank bottoms are not a primary source of contamination at such facilities. A risk-based approach should be employed to evaluate potential sources of release.

- The bill would allow closure of an entire facility in the event that a leak from a single tank is discovered. Shutting down a large terminal is impractical and unnecessary and could seriously affect availability of products. The disadvantages of such action far outweigh any possible advantages.

- Requiring all tank-associated piping to be aboveground is not only impractical and costly, but can potentially create significant safety concerns.

- Requirements that standards be "no less stringent than" those for non-petroleum tanks preempts the regulatory process. Stringency of standards should be determined by EPA as part of a rulemaking process. Such a requirement is arbitrary and fails to recognize the diversity of products stored at aboveground tank facilities. Storage of pesticides and dry-cleaning fluids, for example, may present entirely different concerns than storage of heating oil or other petroleum products.

API members believe that solutions for effective operation of facilities must be risk-based and target actual problems. There is an abundance of evidence demonstrating that government should focus on performance outcomes and let the regulated community find the most efficient and cost-effective method of achieving the desired results. A one-size-fits-all approach wastes resources. For example, the Yorktown Pollution Prevention Study, jointly sponsored by EPA and Amoco, showed that in order to comply with Clean Air Act regulations, Amoco was required to spend $54 million over four years to reduce airborne hydrocarbon emissions from the refinery. However, the Yorktown study showed that Amoco could have reduced those emissions by virtually the same amount for only $10 million--if EPA had the latitude to allow the company to implement scientifically sound alternatives.

It is important for industry and government to work together to achieve mutually understood objectives. API members believe that cooperative processes lead to optimum results. In July 1994, eleven Paw Creek terminal operators in North Carolina signed a pact with state and local environmental officials, designed to address community concerns about operation of the terminal complex, including groundwater remediation. API believes that the innovative process used to develop the Paw Creek pact is an example of the success that can be achieved when government and industry work together to solve difficult problems.

Conclusions

API hopes that this testimony demonstrates the petroleum industry's concern for safe and proper operation of aboveground storage facilities. API members voluntarily participated in the recent survey not only to assess the frequency of groundwater contamination at such facilities, but also to document the improvements that have occurred in recent years and to identify areas for additional improvement. Moreover, in response to concerns about industry practices, API has undertaken an effort to update existing standards and to develop a comprehensive new standard addressing petroleum storage facilities. As a result, API members do not think that additional authority is needed for regulation of aboveground storage tanks and terminals.

---

[1] Standard 653 is referenced by the states of Alaska, Arkansas, Florida, Pennsylvania, Virginia and Washington.

[2] The survey addressed only the tank farm portion of facilities in the three industry sectors. In the transportation sector this involved pipeline bulk storage facilities. Exploration and production facilities were omitted because they are small, remotely located, and are operated differently than large terminals.

[3] The survey results indicate only the presence of contamination; the data do not indicate the volume of contamination at individual sites.

[4] Because of the random nature of the sample, survey results are estimates for the total population of API member company facilities.

[5] There are a variety of reasons why the percent of facilities conducting remediation is not 100% for marketing and transportation facilities with known groundwater contamination. The discovery may have been recently made, the source may be outside the facility, or the source may not have been identified at the time of the survey. Additionally, the contaminant concentration may be low enough that an agency has determined that no action is needed.

[6] When groundwater contamination is identified at a storage facility, it is routinely reported to the appropriate state regulatory agency. After receiving the initial report, the agency involved evaluates the nature of the contamination, determines the urgency or need for further action, and determines the need to oversee any remedial action. Currently, cleanup liability is established by statute in 44 states, and because of broad definitions within many state water quality laws, 41 of those states expressly extend liability to releases to groundwater.

[7] API submitted detailed written comments on H.R. 1360 in a letter dated May 26, 1993, to The Honorable James P. Moran.

Mr. SWIFT. Thank you very much, Mr. DiBona.

Mr. SWIFT. I recognize now Lois Epstein.

STATEMENT OF LOIS N. EPSTEIN

Ms. EPSTEIN. Good morning. My name is Lois Epstein, and I am an engineer with the Environmental Defense Fund, a research and advocacy organization with over 250,000 members nationwide. My background includes extensive work on storage tank issues since 1985, first as a consultant for industry and later as a policy analyst for EDF.

Since early 1988, when the first aboveground tank legislation was introduced on Capitol Hill—this issue has been around for a while—I have assisted House and Senate legislators and State level lawmakers in developing aboveground tank legislation.

Thank you very much, Mr. Chairman, for holding this important hearing today. My comments follow three areas: First, the extent of contamination caused by aboveground tank facilities and resulting environmental and safety hazards; second, what industry and States are doing to address leaking aboveground tank facilities and why these actions are currently insufficient; and third, what authorities EPA lacks to address this national problem and how H.R. 1360 remedies these deficiencies, including how its fee structure addresses the unfunded mandates issue. Please refer to my written testimony for more details.

Let me begin with two remarkable statistics from the recent American Petroleum Institute study. A full 85 percent of monitored refineries and fuel marketing facilities, which are also known as tank farms, show confirmed groundwater contamination. And just as a footnote, the common sense initiative, which I am personally involved in and very supportive of, will only address the refinery tank problem. It will not address the problem of fuel marketing facilities.

The second statistic is that of the aboveground tank facilities with groundwater contamination, at least 27 percent have contamination affecting adjacent property owners, even while a large portion of the facilities don't even know if their contamination has migrated offsite. So despite some governmental oversight of these cleanups, many facilities don't even know if their contamination has affected their neighbors.

Another important fact to keep in mind is that facilities with underground tanks are increasingly replacing these with aboveground tanks. These are smaller aboveground tanks, because of the lack of similar requirements for aboveground tanks.

What are the effects of these leaks and increasing amounts of fuel and other hazardous materials being stored in aboveground tanks? Leaking aboveground tank facilities can pose health or fire hazards in structures such as sewers or basements when gaseous components migrate to enclosed areas and concentrate to toxic or combustible levels. Gasoline from a mobile fuel marketing facility in Brooklyn caused a sewer line explosion in 1950. In 1992, much more recently, several families in Fairfax, Virginia, were forced to leave their homes during cleanup of oil releases from the Star Enterprise facility.

Leaking aboveground tank facilities can adversely affect groundwater and surface water ecosystems, since groundwater supplies approximately 40 percent of surplus water flow nationwide. Tanks, furthermore, can depreciate property values for adjacent property owners when they leak.

As larger quantities of flammable products are stored above ground, there is a greater hazard of fires at storage facilities and not all States have adequate fire codes for smaller aboveground tanks.

What has industry done to prevent leaks and why haven't they done more? The short answer is that petroleum losses are relatively cheap and infrastructure changes may be expensive, particularly over the short term. While the American Petroleum Institute has developed several standards addressing aboveground tank design and operation, the standards are voluntary so existing practices are unlikely to have improved throughout the industry.

Moreover, industry consultants have informed me that there is variability in how well these detailed standards are being followed across the country, and in particular locations depending on who the consultant is that is doing the inspections.

Additionally, an aboveground tank facility owner may wait without fear of civil penalty until State regulators with limited resources get around to overseeing cleanup at a particular site. And the facility owner may only prevent migration at that time. So it is a reactive posture, rather than a preventive, proactive posture.

Even though business forecasters predict that new aboveground tank facility requirements are inevitable, and they are doing this at industry conferences over and over again, many owners are reluctant to invest in infrastructure changes until they are certain about what the new requirements will entail.

Currently, only five States have requirements to prevent aboveground tank facility releases and off-site migration. Thirteen States have some design and operational requirements. Five States have registration requirements and 27 States have no requirements, including the States where our members come from, the members who are here today.

As the media highlight release incidents in particular States, State legislators react by putting requirements in place that will prevent future incidents, as has occurred in South Dakota, Virginia, and potentially will occur in North Carolina.

Less aggressive States wait to see what requirements Congress and EPA enact to eliminate duplicative State-level efforts. There are three deficiencies in EPA's ability to regulate aboveground tank facilities that need to be addressed by legislation. And we have heard about that from EPA today. And these deficiencies are addressed in H.R. 1360, and I think that is what EPA was referring to when they said they need some legislative authorities, and what I think I heard Mr. DiBona support in terms of EPA's increased legislative authority and their needs.

The first deficiency is the authority to prevent underground releases from aboveground tank facilities, including non-oil hazardous substance facilities, through design and operational requirements. Second, EPA needs authority to require that the Federal Government be notified of underground releases. And third, they need some authority to address existing contamination and to ensure that when there is contamination that industry will act to prevent it from migrating offsite.

Other desirable characteristics of this bill include its consolidation of EPA's underground and aboveground tank programs and offices, its fee structure and penalty provisions for noncompliance. H.R 1360 is flexible enough that EPA can design a regulatory program based on the environmental and safety risks posed by different facilities.

Congress also should examine the possibility of utilizing the LUST Trust Fund and the Oil Spill Liability Trust Fund moneys to develop and implement effective aboveground tank programs, including at the State level.

In conclusion, aboveground tank facility releases are pervasive and ongoing. There is no lack of data on this issue. There are more data than when the Reagan administration EPA recommended to Congress that it address underground tanks, and that has been a relatively successful EPA program. It is not one that we have heard challenged for the most part by industry or by States.

The Clinton administration and Congress should ensure passage of similar proactive legislation for aboveground tank facilities to prevent ongoing, not just past releases, and to assist States in developing effective regulatory programs.

Furthermore, legislation will provide certainty for businesses either with aboveground tanks now or considering their purchase as to what the requirements are likely to be.

EDF is anxious to work with Congress, EPA and industry to develop an appropriate legislative and regulatory program that addresses current deficiencies which result in unnecessary underground releases from aboveground tanks and off-site migration from these facilities.

Thank you very much.

Mr. SWIFT. Thank you.

[The prepared statement of Lois N. Epstein follows:]

TESTIMONY OF LOIS N. EPSTEIN, P.E.

My name is Lois Epstein and I am an engineer with the Environmental Defense Fund (EDF), a New York-based, non-profit environmental research and advocacy organization with over 250,000 members nationwide. My background includes extensive work on groundwater protection and storage tank issues since 1985, first as a technical consultant for government and industry and later as a policy analyst for EDF. Since early 1988 when the first aboveground tank legislation was introduced on Capitol Hill, I have assisted House and Senate legislators and state-level lawmakers in developing and promoting aboveground tank legislation.

EDF's comments fall into three areas: first, the extent of contamination caused by aboveground tank facilities and resulting environmental and safety hazards; second, what industry and states are doing to address leaking aboveground tank facilities and why these actions are insufficient, and; third, what authorities and appropriations U.S. EPA currently lacks to address this national problem and how H.R. 1360 remedies these deficiencies.

Background on Underground Releases from Aboveground Tank Facilities

According to 1994 American Petroleum Institute (API) survey data: 1) a full 85% of monitored refineries and fuel marketing facilities (or "tank farms") show confirmed groundwater contamination,[1] and; 2) of the refining, marketing, and transportation-related aboveground tank facilities with groundwater contamination, at least 27% have contamination affecting adjacent property-owners, even while a large proportion of surveyed facilities do not know if their contamination has migrated off-site.[2] The first API statistic was independently confirmed in Virginia using groundwater characterization studies for facilities with greater than one million gallons of storage capacity, with approximately 85% of the facilities showing contamination.[3] Similarly, in New York, 80% of the aboveground tank facilities with over 400,000 gallons of storage capacity have confirmed contamination.[4]

To place these numbers in perspective, there are approximately 180 operating refineries in the U.S., approximately 1,200 fuel marketing facilities (not including service stations), and approximately 2,000 transportation-related aboveground tank facilities. Also, chemical industry facilities have approximately 200,000 aboveground tanks, and non-oil industries store bulk fuel in 100,000-200,000 tanks. Attachment A lists some of the largest known underground releases from aboveground tank facilities in the U.S.

Another important background fact is that facilities with underground tanks for storage are increasingly replacing them with aboveground tanks because of existing, national underground tank regulatory requirements and the lack of similar requirements for aboveground tanks. A Steel Tank Institute survey of 200 retail petroleum marketing, government/military, industrial/commerical processing, commercial fleet fuel storage, and commercial airports and chemical plant facilities found that most facility owners or operators planned to buy an aboveground tank for their next tank purchase -- over 50% said they would purchase an aboveground tank, compared to less than 20% choosing an underground tank.[5] Grace Specialty Chemicals, a New York-based company with more than 90 plants across the U.S. reported in 1990 that it was replacing its 274 underground tanks with aboveground tanks.[6]

What are the effects of aboveground tank facility leaks and increasing amounts of fuel and other hazardous materials stored in aboveground tanks? Leaking aboveground tank facilities can pose health or fire hazards in structures such as basements when gaseous components migrate into enclosed areas and concentrate to toxic or combustible levels. Gasoline from a Mobil fuel marketing facility in Brooklyn, New York caused a sewer line explosion in 1950. More recently, in 1992 several families in Fairfax County, Virginia were forced to leave their homes during cleanup of oil releases from the nearby Star Enterprise aboveground tank facility.

Leaking aboveground tank facilities also may adversely affect groundwater and surface water ecosystems, since groundwater supplies approximately 40% of surface water flow nationwide. In Port Everglades, Florida, underground contamination from aboveground tank facilities flowed for years into the Intracoastal Waterway, a surface water body bordering many protected natural areas. In East Providence, Rhode Island, a major underground oil release is currently migrating to the Runnins River.

Leaking aboveground tank facilities frequently depreciate property values for nearby property owners. In Fairfax County, VA, as many as 450 families will be compensated approximately $150 million dollars for the drop in value of their homes since the Star Enterprise release was discovered.[7]

As larger quantities of flammable products are stored aboveground, there is a correspondingly greater likelihood of fires at storage facilities, and not all states have adequate fire code requirements for small aboveground tanks. Additionally, there is a need for state-level enforcement to ensure that codes are implemented adequately at the local level.

Industry and State Responses to Underground Releases from Aboveground Tank Facilities

Why doesn't industry act to prevent product leaks? The short answer is that petroleum losses are relatively cheap and infrastructure changes may be expensive, particularly over the short-term. The 1994 API survey shows that a relatively small percentage of facilities employ existing technologies to prevent or reduce groundwater contamination (see Table, below).[8]

Percent of Aboveground Tank Facilities Employing Improvements to Reduce Groundwater Contamination

Type of Improvement	Type of Facility		
	Refining	Marketing	Transportation
Release prevention barriers underneath tank bottoms	12%	17%	20%
Corrosion protection for tank bottoms	45%	70%	81%
Overfill protection for tanks	64%	82%	88%
Corrosion protection for all buried piping	15%	60%	96%
Aboveground piping (75% or more at the facility)	78%	54%	11%

In fact, aboveground tank facility owners may wait, without fear of civil penalties, until state regulators with limited resources decide to oversee a cleanup effort at their site, and only act to prevent migration at that time. Additionally, even though business forecasters uniformly predict that new aboveground tank facility upgrading requirements are inevitable, many are reluctant to invest in infrastructure changes until they are more certain about what the new requirements will entail.

While API has developed several standards addressing aboveground tank design and operation, e.g., API 650 and 653, these standards are voluntary so existing practices are unlikely to have improved uniformly throughout the industry. Additionally, consultants have informed EDF that there is extreme variability in how well API's rather detailed standards are followed. Nevertheless, these standards are valuable baselines, and EPA and states should incorporate them, as appropriate, into their regulatory programs. In reality, certain companies are performing necessary aboveground tank facility upgrades, but there are many industry laggards (note that API member companies represent only about half of operating refineries).

The 1994 API survey cited above did not provide useful information on the extent of remediation activities occurring at aboveground tank facilities, particularly prevention of off-site migration. As discussed in EDF's analysis of the survey's report,[9] it is not true (as API states) that "in virtually all cases where contamination exists, remedial activities are taking place"[10] -- in fact, EDF contends that in very few cases are remedial activities occurring, but API nevertheless makes this statement because it classifies "monitoring" as a remedial activity in its survey report. Additionally, the survey's report mischaracterizes the extent to which federal, state, or local agencies oversee, review, or supervise remedial activities -- if governmental agencies were overseeing remedial activities at most petroleum storage facilities, it is extremely unlikely that 20% of the fuel marketing facilities with known groundwater contamination would answer that they "Don't Know" if their contamination had migrated off-site.

Currently, five states have extensive requirements to prevent aboveground tank facility releases and off-site migration, thirteen states have some design and operational requirements, five states have only registration requirements, and twenty seven states have no requirements (see Attachment B). As the media highlight aboveground tank facility release incidents in particular states, state legislators react by putting requirements in place that would prevent future incidents, as has occurred in South Dakota, Virginia, and potentially in North Carolina. Less aggressive states generally choose to wait and see what requirements Congress and EPA will enact, to eliminate duplicative state-level efforts.

**EPA Currently Lacks Authority and Appropriations to Address
Underground Releases from Aboveground Tank Facilities**

There are three major deficiencies in EPA's ability to regulate aboveground tank facilities that need to be addressed by federal legislation and appropriations, as in H.R. 1360:

* authority to prevent underground releases from aboveground tank facilities, including non-oil hazardous substance storage facilities, through design and operational requirements;

* authority to require the federal government's National Response Center to be notified of underground releases, and;

* authority to address existing contamination and ensure no off-site migration.

Other desirable characteristics of H.R. 1360 include its consolidation of EPA's underground and aboveground tank programs, its penalty provisions for non-compliance, its fee provisions to assist in implementation, and the General Accounting Office study of transportation-related tanks and piping to determine if they are adequately regulated by the U.S. Department of Transportation for the purpose of environmental protection. H.R. 1360 is also flexible enough that EPA can design a regulatory program based on the environmental and safety risks posed by individual facilities.

To prevent releases from aboveground tank facilities, EPA must have the authority, as in H.R. 1360, to require:

* state-of-the-art technology such as a double-bottom with interstitial leak detection for new tanks and for existing tanks within a reasonable timeframe (depending on site specific conditions and tank usage, temporary tanks such as production tanks, may not require this type of containment);[11]

* corrosion protection for tank bottoms and buried piping;

* spill and overfill prevention technology and containment of releases during product transfers;

* moving piping aboveground to the maximum extent feasible;

* tank and piping installation and testing requirements;

* facility inspection requirements,

* tank closure requirements; and,

* enforcement provisions for violation of release prevention standards, including penalties and citizen suit provisions.

Desirable changes to H.R. 1360 include language that:

* requires EPA to consolidate its aboveground tank-related responsibilities in one office,

* requires EPA to base its aboveground tank facility regulations on the risk of releases and accidents from particular types of facilities or types of tanks and piping,

* clarifies how the legislation relates to Section 311 under the Clean Water Act which covers prevention and cleanup of releases from aboveground tank facilities to surface waters. The current lack of federal resources devoted to the Section 311 Spill Prevention, Control, and Countermeasures program has, however, greatly limited this program's effectiveness,

* allows EPA to levy penalties for oil and hazardous substance releases,

  *   streamlines the facility notification and inspection provisions, and

  *   permits access to the LUST Trust Fund for aboveground tank programs in a manner similar to that used by EPA and state underground storage tank programs.[12] Congress also should explore utilizing funds from the Oil Spill Liability Trust Fund, established under the Oil Pollution Act of 1990, to help develop and implement an effective aboveground tank program.

Many of these recommendations address EPA's concerns with H.R. 1360, as stated in the Agency's May 2, 1994 letter to Chairman Swift.

In conclusion, aboveground tank facility releases are clearly pervasive and ongoing. There is no lack of data on this issue -- in fact there are more data than when the Reagan Administration EPA recommended to Congress that it address underground tanks during the 1984 Resource Conservation and Recovery Act reauthorization process. The Clinton Administration and Congress should ensure passage of similar legislation for aboveground tank facilities to prevent ongoing releases, to assist states in developing effective regulatory programs, and to provide certainty for businesses either with aboveground tanks now or considering their purchase. EDF is anxious to work with Congress, EPA, and industry to develop an appropriate legislative and regulatory program that addresses current deficiencies which result in unnecessary underground releases and off-site migration from aboveground tank facilities.

---

[1] American Petroleum Institute, "A Survey of API Members' Aboveground Storage Tank Facilities," Washington, DC, July 1994, Table III, p. 15.

[2] Ibid., p. A12.

[3] "Virginia Groundwater Studies: 40% More Contamination," Aboveground Tank Update, Volume 5, Number 1, January 1994, p. 10.

[4] "Highlights of Forums on Aboveground Oil Storage Facilities," U.S. Environmental Protection Agency Aboveground Oil Storage Facilities Workgroup, November 24, 1993, p. 2.

[5] "Aboveground Storage Tank Boom -- Will the Trend Continue?", survey conducted by Product Evaluation Inc., Chicago, IL for the Steel Tank Institute, Lake Zurich, IL, January 1993, p. 1.

[6] "Grace Yanks Tanks Early," Chemicalweek, Chemical Week Associates, New York, NY, Vol. 146, No. 18, May 9, 1990, p. 36.

[7] Steve Bates, "Star to Pay $50 Million in Oil Leak: Officials Outline Plan to Compensate Affected Va. Families," Washington Post, September 5, 1992, p. 1.

[8] American Petroleum Institute, op. cit., pp. A9-10.

[9] Lois N. Epstein, "A Survey of API Members' Aboveground Storage Tank Facilities (July 1994): A Critical Analysis," Environmental Defense Fund, July 18, 1994, pp. 2-3.

[10] American Petroleum Institute, op. cit., p. iii.

[11] EPA already has this authority under Section 4113 of the Oil Pollution Act of 1990. Under this law, EPA's determination of whether liners should be utilized underneath aboveground tanks should have been sent to Congress by August 1991, with implementation of the recommendations six months later.

[12] Since its inception, the LUST Trust Fund established by Section 205 of the Superfund Amendments and Reauthorization Act of 1986 has not been fully appropriated to EPA for the purpose for which it was set up with a tax at gasoline pumps, i.e., for leaking underground storage tank cleanup activities.

Attachment A

Summary of Major (Above 100,000 Gallons) Aboveground Tank
Facility Releases of Petroleum Products Underground

<u>Location</u>	<u>Facility Type</u>
Anchorage, Alaska	Air Force Tank Farm
El Segundo, California	Refinery
Martinez, California	Refinery
Torrance, California	Refinery
Deerfield Beach, Florida	Oil Recycler/Storage Facility
Port Everglades, Florida	Transportation/Marketing
Hartford, Illinois	Refinery
Granger, Indiana	Marketing
Whiting, Indiana	Refinery
Cattlettsburg, Kentucky	Refinery
Plaquemine, Louisiana	Chemical Plant
Sparks, Nevada	Transportation
Paulsboro, New Jersey	Marketing
Brooklyn, New York	Marketing
Syracuse, New York	Marketing
Greensboro, North Carolina	Marketing
Ponca City, Oklahoma	Refinery
Tulsa, Oklahoma	Refinery
Philadelphia, Pennsylvania	Refinery
East Providence, Rhode Island	Marketing
Spartanburg, South Carolina	Marketing
Austin, Texas	Marketing
Fairfax, Virginia	Marketing
Seattle, Washington	Refinery
Tacoma, Washington	Marketing

Attachment B

Aboveground Tank Facility Requirements by State as of 9/94

No Req'ts.	Only Registration Req'ts.	Some Req'ts.	Extensive Req'ts.
Alabama	Arkansas	Alaska	Florida
Arizona	California	Colorado	New York
Connecticut	Kansas	Delaware	Rhode Island
Georgia	Nevada	Louisiana	South Dakota
Hawaii	North Carolina	Maryland	Virginia
Idaho		Massachusetts	
Illinois		Minnesota	
Indiana		New Jersey	
Iowa		Oklahoma	
Kentucky		Pennsylvania	
Maine		Texas	
Michigan		Washington	
Mississippi		Wisconsin	
Missouri			
Montana			
Nebraska			
New Hampshire			
New Mexico			
North Dakota			
Ohio			
Oregon			
South Carolina			
Tennessee			
Utah			
Vermont			
West Virginia			
Wyoming			

Mr. SWIFT. Mr. Clark Houghton.

STATEMENT OF CLARK HOUGHTON

Mr. HOUGHTON. My name is Clark Houghton, and I am president of the Missouri Petroleum Marketers Association. I am here today representing the Petroleum Marketers Association of America. I also happen to be an engineer. I do have 42 years in this business.

PMAA is a federation of 43 State and regional associations which represent more than 10,000 small, independent petroleum marketers. Collectively, these marketers sell more than 45 percent of the gasoline and 60 percent of the diesel fuel and 75 percent of the home heating fuel.

We commend the chairman for holding these hearings on this important topic and aboveground storage tank leaks. Let me say from the outset we are against the bill. The problem with this legislation is that it attempts to impose a comprehensive regulatory solution to a problem that may not exist or is already covered by existing regulations.

The truth of the matter is simply not enough current information currently is available to define the AST universe, let alone suggest any significant problems with AST's. Now, to my knowledge, which is very limited I must admit as I live in the heart of Missouri, there has never been a comprehensive study to gather this data. I hear EPA has data. I hear that EPA is studying it. Being—like I say, I am a little confused as to what is comprehensive.

So to relate to the location of the potential harm to the waters of the United States, nor is there data available indicating how many existing AST's are used exclusively for petroleum products, what kind of tanks are leaking, field directed or plant fabricated, how many are leaking, and how much product has been lost.

There have been two major notable events in the past couple of years, which have been mentioned. I do think that the regulation that came out in July of this year requiring certain facilities to have more than just a simple paper cutter or word processor, the SPCC plan would help identify this problem and prevent great harm there. Would have reduced the harm, not prevented it; reduced it.

And I believe that existing technology reporting and response regulations on both the Federal and State level are more adequate to ensure that AST's operate safely. The Administrator of the EPA seems to think so also as she came out against this legislation this summer. Now I hear that they would like some other legislation.

Another important reason is that at this time there is an economic burden that it would place on small, independent owners and operators. And 89 percent of PMAA's members represent small businessmen and women.

I am an operator of an 8 million gallon a year oil distributing business. That is a small business. We employ a maximum of 50 people in convenience stores in tank wagon delivery and transport delivery. We are beginning to suffer under all the regulations. More than beginning to suffer.

Most of these operations are family owned and operated and passed down from generation to generation. PMAA members do not own or operate the large tank farms. That is obvious. More typically their facilities consist of one, two, or maybe four AST's, smaller holding capacity, 20,000 gallons per tank or less, and are not found on large farms.

They are already complying with extensive State and Federal regulations. These include SPCC plans, engineering studies, site assessments, employee certification and training, inspection, reporting, containment, and response activities, and other expenses.

The expense of a small marketer cleaning a tank bottom for inspection is very large. To remove two 55-gallon drums of water, contaminated water, pumped out of the bottom of an underground tank of one of my facilities is going to cost approximately $2,000 to dispose of.

That is because you have to have a consultant, and I am not qualified as I am not an impartial consultant in my company, to analyze the product and take care of getting rid of it. Two thousand dollars to get rid of two 55-gallon drums of water contaminated with some gasoline in gasoline storage tanks. Just imagine what it would be to remove the bottoms where the pipe is 4 to 5 inches or more from the bottom of the tank on a vertical tank. You are talking 300, 400, 500 gallons. Maybe more.

So I just want to say that it is a very expensive process, and it should be considered.

Some of the regulations in the past have disrupted the delivery of petroleum products to the rural areas they serve, and it is certainly not the intent of Congress or the sponsors of this bill that such a result would follow such an action.

Before any legislation such as H.R. 1360 is considered, more data needs to be collected, which is evident from this hearing. Existing enforcement and regulation should be reviewed for adequacy and a cost-benefit analysis be done on the effect that this legislation would have on the small, independent marketer. And I urge this committee to reject this legislation until the extent of the problem sees the full light of day.

Thank you, Mr. Chairman, and members of the committee.

[The prepared statement of Clark Houghton follows:]

STATEMENT OF CLARK HOUGHTON, STATE EXECUTIVE, PETROLEUM MARKETERS OF AMERICA

Good morning Mr. Chairman and members of the subcommittee. My name is Clark Houghton, president of the Missouri Petroleum Marketers Association and director of Mid-Missouri Oil Co. I have served as regional vice president of the Petroleum Marketers Association of America (PMAA), and chaired that organizations Operations and Engineering Committee for 6 years. I hold an engineering degree from the University of Missouri at Rolla and consider myself knowledgeable in the field of aboveground storage tank engineering.

I am appearing today in my capacity as engineering consultant for PMAA. PMAA is a federation of 43 state and regional trade associations representing more than 10,000 small and independent petroleum marketers. Collectively, these marketers sell more than 45 percent of the gasoline, 60 percent of the diesel fuel and 75 percent of the home heating fuel consumed annually in the United States.

PMAA commends the chairman for holding these hearings on the very important topic of aboveground storage tank legislation. Let me say from the outset, that PMAA does not support H.R. 1360. The problem with this legislation, is that attempts to impose a comprehensive regulatory solution to a problem that may not even exist. The truth of the matter is that there is simply not enough information currently available that defines the AST universe, let alone that suggests any significant problems with AST's exist.

For example, there has never been a comprehensive study, to my knowledge, that has gathered such basic data as the number, age, size and location of AST tanks and facilities existing throughout the country. Nor is there any data available that would indicate how many of the existing AST's are used exclusively for petroleum products, which tanks are leaking or have leaked, and how much product has been lost.

There have been in the recent past, one or two notable AST events that have caused significant environmental harm. In Fairfax County, Virginia, an AST failed and leaked thousand of gallons of gasoline, contaminating the environment and requiring a residential evacuation. Truthfully, it was a mess and I would imagine the reason why Congress is considering H.R. 1360 today. However, there are serious questions as to whether the owner reported the leak at the outset when something could have been done about it, thus exacerbating the problem. That particular leak could have been mitigated considerably by acting quickly to remove the product from the tank when the leak was discovered. Federal law requires this type of response action. In my view, had the operator followed existing Federal requirements for reporting and responding to such leaks, the amount of product spilled and the environmental damage caused by the leak would have been far, far, less than what actually happened. Consequently, it is probably more appropriate to look at existing enforcement measures rather than imposing a new and expansive regulatory burden on all AST owners and operators.

In fact, I believe that existing technical, reporting and response regulations on both the Federal and State level, are more than adequate to ensure that AST's operate safely. The administrator of the Environmental Protection Agency thinks so too. That's why she came out against this legislation earlier this summer.

Another important reason this legislation is a bad idea at this time is the economic burden it would place on the small independent owner and operators. Eighty-nine of PMAA's members represent small business men and women. More often than not, the operation is family owned and operated and is passed down from generation to generation. PMAA's members do not own or operate the large tank farms such as the one you see down on Route 95, south of Washington. More typically their facilities consist of one or two AST's and much smaller holding capacity than those found on large tank farms. Moreover, their business, are run like all small businesses, on a limited operating margin.

These operators are already complying with expensive State and Federal regulations. Engineering studies, site assessments, employee certification and training, inspection, reporting, containment and response activities represent a huge expense for small petroleum marketers. The additional expense imposed by H.R. 1360 would force many of PMAA's members out of business and disrupt the delivery of petroleum products to the rural areas they serve. Certainly it is not the intention of Congress, or the sponsors of this bill, that such a result would follow from enactment.

I would like to summarize by saying that before any legislation such as H.R. 1360 is considered, more data needs to be collected, existing enforcement and regulations should be reviewed for adequacy and cost benefit analysis be done on the effect this legislation would have on the small independent petroleum marketer. I urge this committee to reject this legislation until the extent of the problem sees the full light of day. Thank you Mr. Chairman and members of the committee for giving me this opportunity to address you on this very important subject.

Mr. SWIFT. Thank you, Mr. Houghton. And forgive me for mispronouncing your name.

Mr. HOUGHTON. Oh, that is all right. I have been called a lot worse than that.

Mr. SWIFT. Marshall T. Mott-Smith.

STATEMENT OF MARSHALL T. MOTT-SMITH

Mr. MOTT-SMITH. Good morning, Mr. Chairman, and members of the subcommittee. The Florida Department of Environmental Protection appreciates the opportunity to testify before the subcommittee on the important issue of aboveground storage tank regulation.

My name is Marshall Mott-Smith, and I have been with Florida's environmental agency for 17 years. And I have been administrator of the storage tank program since 1986.

Florida has been regulating underground and aboveground storage tank systems since 1983. Because the State relies on groundwater for over 95 percent of its drinking water, it has regulations that are more stringent than EPA's for preventing and cleaning up contamination from storage tank systems.

In 1983, the Florida legislature passed laws that gave the Department authority to adopt rules for underground and aboveground tanks with storage capacities greater than 550 gallons that contained petroleum products and other pollutants. Legislature chose to regulate all storage tank systems because aboveground tanks pose the same risk to State ground and surface waters as underground tanks.

The State adopted its first storage tank rules in 1984 and later revised them in 1991, to require secondary containment for all new aboveground storage tanks and integral piping in contact with the soil.

We are currently proposing additional rule changes and updating and adding relevant industry standards, including the new API Standard 2610.

Program statistics for the State's aboveground storage tank program are attached in my appendices of my written statement. Florida has approximately 28,000 active aboveground tanks located at 11,500 facilities. Four thousand six hundred of these facilities have underground tanks in addition to the aboveground tanks on site.

Fifty-four percent of the State's active tanks are under 1,100 gallons. There are 1,500 bulk product tanks and 96 percent of all product storage capacity in Florida is contained in these tanks.

Over 22,000 discharges have been reported since the program began. Now that comes from both underground and aboveground tanks. However, it is often hard to determine the cause of a leak because many owners specify the cause as unknown when they file a discharge report with the department.

Because aboveground tanks are co-located with underground tanks at approximately 40 percent of Florida's registered facilities, our statistics do not provide a clear picture of the number of leaks from aboveground tanks. The percentage of open facilities with aboveground tanks reporting releases is 22 percent; however, 66 percent of the bulk storage facilities and 54 percent of the retail stations have reported a release.

The average cost of cleaning up a petroleum-contaminated site in Florida is $275,000. Florida's Inland Protection Trust Fund generates $160 million a year and approximately $400 million has been spent for site cleanup costs to date.

Most of the sites receiving money from the trust fund have underground storage tank systems. Nevertheless, the most costly sites are those with aboveground bulk storage tanks. The cost of cleaning up the Port Everglades terminal port facility near Ft. Lauderdale to meet State water quality standards could be as much as $50 million. The Port Everglades facility has 264 aboveground tanks, has been in operation since 1928, and currently ranks as one of the largest ports in the Nation.

Reported releases from tank spills, pipeline leaks, and other sources date back to the 1950's and several areas have over 5 feet of free product thickness floating on the groundwater table. One area has almost 8 feet of free product. An investigation is under way to comprehensively address the problems and develop a site remedial action plan.

Federal facilities are another example of problems associated with aboveground tanks. Most of the facilities at Florida Air Force and Navy bases have reported leaks from their aboveground tank systems. Due to the large number of contaminated sites, base closures, and shrinking defense budgets, it has been difficult for many military installations to obtain funding for upgrading existing tanks and to clean up contaminated sites. Many of the military bases have soil-covered, concrete cut and cover tanks from the 1940's that are similar to an aboveground field-erected steel tank in construction and operation.

The U.S. Navy Cecil Field installation near Jacksonville has had numerous petroleum releases, particularly from their cut and cover tanks. One discharge was caused by vandalism but nevertheless it allowed 1 million gallons to surface in groundwaters nearby. Secondary containment would have prevented any groundwater contamination from the incident.

To verify the compliance status of registered tanks, our department developed a compliance inspection program with local governments to perform inspections of all facilities. Over 130,000 inspections have been performed to date. That is both UST and AST.

Although most of the inspections have been with underground tanks, department and county inspectors have performed over 28,000 inspections at facilities with only aboveground storage tanks. The program has resulted in a significant increase in the compliance rate of all storage tank facilities. For aboveground tank facilities, the rate of compliance has increased from 15 percent in 1988 to 65 percent in 1993.

A Federal program for the regulation of aboveground storage tank systems is needed. EPA has regulations for underground storage tanks but does not have regulations that protect groundwater resources from aboveground storage tank system leaks. A leak from an aboveground storage tank system will cause the same environmental damage as a leak from an underground storage tank system. Aboveground bulk storage tanks especially should be regulated due to their size and extensive potential to cause environmental damage.

Florida's experience and statistics and recent incidents in Pennsylvania and Virginia provide ample justification for the regulation of these tanks. The growing number of States developing their own regulations highlights the need for Federal regulations.

Also, Florida and many other States have oil spill prevention programs that regulate aboveground tanks at coastal terminals. Florida has recently combined its two programs to reduce the costs of AST regulation. Consequently, any Federal aboveground tank program should be combined with the EPA underground program and the oil spill prevention program.

A national regulatory program for aboveground tanks is needed to better protect the health and safety of U.S. citizens and the quality of ground and surface water; to provide consistent national standards for aboveground storage tanks and piping construction, release detection, operation, and maintenance to prevent the release of petroleum products and hazardous substances; coordinate the oil spill prevention and groundwater protection efforts of Federal and State agencies; to prevent future releases of petroleum products that could result in costly cleanup programs at public expense; and to ease regulatory confusion for storage tank owners that must comply with many different State and Federal regulations and agencies.

In conclusion, the Florida Department of Environmental Protection supports legislation that would establish a Federal program for the regulation of aboveground storage tank systems. A program similar to the very successful EPA underground storage tank program that provides financial and technical support to the States while allowing flexibility for State-specific needs is recommended. Thank you.

Mr. SWIFT. Thank you, very much. I recognize now Mr. Anthony O'Neill.

STATEMENT OF ANTHONY R. O'NEILL

Mr. O'NEILL. Good morning, Mr. Chairman. I am Tony O'Neill, vice president, government affairs of the National Fire Protection Association, and it is my pleasure to present NFPA's position on H.R. 1360.

The National Fire Protection Association, which represents some 60,000 members, is dedicated to reducing life loss and property damage from destructive fire, and we are pleased to support H.R. 1360, the Safe Aboveground Storage Tank Act of 1993. And in particular we are pleased to support the incentives that are built into the legislation to safeguard aboveground storage tanks from destructive fires.

Recognizing that this is primarily an environmental bill, Congressman Moran is to be commended for providing incentives for improved fire safety by specifying the adoption and proper enforcement of state-of-the-art national consensus fire codes such as NFPA 30 Flammable and Combustible Liquids Code and the Automobile and Marine Service Station Code, and the other model uniform fire codes that are used at the State and local level. Although these are intended primarily for fire safety, they do indeed specify important environmental protection safeguards.

Because H.R. 1360 recognizes that the fire safety of flammable liquid storage tanks is generally the responsibility of State and local governments, the legislation properly refers to various model fire codes and channels a portion of the fees collected under section 4 to training programs for fire code inspection personnel to fund costs of the agency in administering the notification program and, as importantly, to fund certain research activities.

Now, why are these modern fire safety codes so important? Aboveground storage tanks, AST's, for flammable and combustible liquids, such as gasoline and fuel oil, can represent a major fire hazard to our citizens, to firefighters, and indeed to communities, property and businesses, whenever the tanks are not properly engineered and protected in accordance with the modern fire codes.

To demonstrate the severe fire problems especially to firefighters when tanks are not properly designed and installed, we have enclosed examples and specifically the published reports on two incidents that caused the death of eight firefighters and in addition dozens of firefighter injuries in Gadsden, Alabama, in 1976, and Kansas City, Kansas, in 1959.

These incidents, and similar cases, illustrate the need for the following basic safeguards, all of which have been included in the modern fire codes over the years: Overfill protection; spill control (diking and containment); emergency venting; and minimum siting requirements.

I have also provided for the subcommittee as part of our written testimony statistics and case histories on tank farm fires in the United States, service station fires, and those situations, and also refinery fires from the national statistical bases that we work with in conjunction with Federal Government agencies.

Now, in our written statement we provide examples of what can happen when these basic fire protection requirements are not included and not managed into the tank design and construction. I won't get into those details right now in this oral statement, but they are contained in the written statement.

And one point that needs to be emphasized, I think, is why is there renewed interest in the safety of aboveground storage tanks?

For over 100 years, we have been developing the automobile and in that period of time it has been accompanied with the need to supply gasoline, often at remote locations; thus, in the early days the advent of aboveground storage tanks, many of which are still in service. But over the years many of these tanks became subject to code requirements for emergency venting, spill control, overfill prevention, et cetera, but not before some of the major incidents that I referred to.

During the evolution of the gasoline marketing and distribution industry, more and more of these storage tanks were put underground where they are relatively safe from fire. Rarely do you see today aboveground storage tanks at service stations in urban and suburban areas.

In recent years, there has been a movement by gasoline marketers to once again utilize aboveground storage tanks at service stations for a variety of reasons, including some of the testimony that you have heard today; the liability exposure, the cleanup costs associating with leaking underground storage tanks, and the ability to visually inspect these aboveground storage tanks. So the code groups, including those of the National Fire Protection Association, have been responding to the fact that we anticipate that we are going to see more aboveground storage tanks in our future.

And the current editions of our codes are reviewing such issues as design and construction requirements so that tanks will stay together, will hold liquid without mechanical failure, spill control requirements. The concept of secondary containment is being studied. The code requires any tank taken out of service must be emptied, cleaned and rendered vapor free. The code has special requirements for corrosion protection on underground tanks and piping, and special provisions for flood-prone areas as well as requirements to prevent overfilling of the tanks.

In conclusion, Mr. Chairman, NFPA supports H.R. 1360 because it recognizes the fire hazards of aboveground storage tanks, provides incentives to ensure that these tanks are properly protected through the adoption and enforcement of nationally recognized model fire safety codes at the State and local level, and allocates a portion of the funds collected in order to help fund these activities.

Mr. Chairman, I would be pleased to answer any questions.

Mr. SWIFT. Thank you, Mr. O'Neill.

[The prepared statement of Anthony R. O'Neill follows:]

STATEMENT
of the
NATIONAL FIRE PROTECTION ASSOCIATION

Good morning, Mr. Chairman, I am Tony O'Neill, Vice President --Government Affairs of the National Fire Protection Association and it is my pleasure to present NFPA's position on HR 1360. I am accompanied by Bob Benedetti, NFPA's Flammable Liquids Engineer who is available to address any technical questions about our codes that are referenced in HR 1360.

The National Fire Protection Association (NFPA), representing some 60,000 members and dedicated to reducing life loss and property damage from destructive fire is pleased to support HR 1360, "The Safe Aboveground Storage Tank Act of 1993," and in particular, the incentives built into the legislation to safeguard aboveground storage tanks from destructive fire.

Recognizing that this is primarily an environmental bill, Congressman Moran is to be commended for providing incentives for improved fire safety by specifying the adoption and proper enforcement of state-of-the-art national consensus fire codes such as NFPA 30 "Flammable and Combustible Liquids Code" and the "Automotive and Marine Service Station Code," NFPA 30A. As we point out later in this statement, NFPA 30 and 30A, although intended primarily for fire safety, do indeed specify important environmental protection safeguards. (Copies of these codes have been provided to the Subcommittee).

Because HR 1360 recognizes that the fire safety of flammable liquid storage tanks is generally the responsibility of state and local governments, the legislation properly refers to various model fire codes and channels a portion of the fees collected, under Section 4 of the Act, "Notification," to training programs for fire code inspection personnel, to fund costs of the agency in administering the notification program, and to fund certain research activities.

WHY ARE MODERN FIRE CODES SO IMPORTANT?

Simply stated, aboveground storage tanks (AST's) for flammable and combustible liquids (such as gasoline and fuel oil) can represent a major fire hazard to our citizens, to firefighters, and, indeed, to communities, property and businesses, ...**whenever** the tanks are not properly engineered and protected in accordance with modern fire codes.

To demonstrate the severe fire problems, especially to firefighters, when tanks are not properly designed and installed, we have enclosed as appendices to this statement, published reports on two incidents that caused the death of 8 firefighters and, in addition, dozens of firefighter injuries (Gadsden, Alabama, August 31, 1976 and Kansas City, Kansas, August 18, 1959).

These incidents and similar cases illustrate the need for the following basic safeguards, all of which have been included in modern fire codes over the years as a result of these and other incidents:
- Overfill protection
- Spill control (diking and containment)
- Emergency venting
- Minimize siting requirements

as well as specially designed normal vent piping, adequate supports and anchoring, and basic engineering design and construction requirements for the tank itself.

WHAT CAN HAPPEN WHEN BASIC FIRE SAFETY REQUIREMENTS ARE LACKING?

Here is a an example of fire that can occur when basic safeguards at a service station are lacking; an aboveground gasoline storage tank (AST) is overfilled by the delivery truck or, a connecting pipe is broken by an automobile accident, and gasoline is flowing on the ground to an ignition source (could be the delivery truck, a heater, someone smoking, etc.). Fire then flashes back to the tank causing a huge spill fire that spreads to adjacent properties and/or a pool fire that starts impinging on the AST, heating the tank like a tea kettle. Pressure builds up and, if there is inadequate emergency venting, the tank will rupture, spilling burning gasoline over a wide area and, worse, can explode violently in a phenomenon known as a "BLEVE" or boiling liquid expanding vapor explosion. Unlike the tea kettle where water in the form of steam ruptures the tank, a BLEVE allows flaming gasoline or other flammable liquids to spread rapidly, almost instantaneously, over a widespread area.

Fortunately, there is now general recognition of, and therefore acceptance of, the fire safety principles contained in modern fire codes so that very few incidents occur such as Gadsden, Alabama and Kansas City, Kansas. However, as shown in the Gadsden case, there is still the potential for someone to padlock the emergency vent closed or to subvert other basic fire safety requirements. This is why an active on-going code enforcement and inspection program is so important for aboveground flammable liquids storage tanks.

WHY THE RENEWED INTEREST IN THE SAFETY OF ABOVEGROUND STORAGE TANKS?

The nearly 100 year history of motorized vehicular transportation in America has been accompanied with a need to supply the automobile with gasoline, often at remote locations; thus in the early days the advent of aboveground storage tanks, many of which are still in service. Over the years, many of these tanks became subject to code requirements for emergency venting, spill control, overfill prevention, etc. but not before major firefighter and civilian tragedies occurred, such as Gadsden, Alabama and Kansas City cited above.

During the evolution of the gasoline marketing and distribution industry, more and more storage tanks were put underground where they are relatively safe from fire. Rarely did you see aboveground storage tanks at service stations in urban and suburban areas.

A review of fire experience in service stations, refineries, or tank farms (attached as appendices to this testimony) demonstrates that most reported fires in this group occur at service stations, but some of the most severe fires have occurred at refineries and tank farms. This is what one would expect with below ground tanks at service stations and aboveground, larger tanks at refineries and tank farms. These incidents also suggest that the most severe fires typically begin near the tank. NFPA's analysis shows that in recent years only about 5% of service station fires have begun in the area of the tank or any other product storage area.

In recent years, there has been a movement by gasoline marketers to once again utilize aboveground storage tanks (AST's) at service stations for a variety of reasons; including the liability exposure and clean-up costs associated with leaking underground storage tanks, and the ability to visually inspect aboveground storage tanks. These factors are offset by space limitations and other considerations at service stations. Nonetheless, given the probability that there is going to be an increase in the number of AST's, fire safety experts especially those in the fire services are concerned that the lessons of the past may be forgotten.

RESPONSE OF THE NFPA FLAMMABLE AND COMBUSTIBLE LIQUIDS CODE COMMITTEES TO ENVIRONMENTAL NEEDS

Current editions of NFPA 30 and 30A (referenced above) reflect a commitment by these national consensus code development committees to recognize both the traditional fire safety requirements of AST's and the need for enhanced environmental protection measures as follows:

- Design and construction requirements for tanks ensure that the tank is strong enough to hold the liquid without mechanical failure and that the liquid will not corrode the tank shell.

- Spill control requirements include remote impounding basins or a dike around the tank(s) or a combination of both and these measures are intended to intercept and retain any liquid released from an AST as a result of overfilling or breaking of a pipe connected to it.

- The concept of secondary containment originated with underground tanks because of corrosion problems and is also recognized in the codes but is not always spill control; particularly if a pipe break or overfill spills liquids directly to the ground.

- The code requires any tank taken out of service must be emptied, cleaned and rendered vapor-free as both a fire/explosion prevention measure and an important environmental consideration.

- The code has special requirements for corrosion protection on underground tanks and piping, adequate support to prevent uneven settling and the supports for AST must be resistant to failure from fire exposure.

- There are special requirements for tanks in flood-prone areas to prevent movement or shifting of the tanks.

- The code contains requirements to prevent overfilling the tank which is both a fire safety and environmental protection measure.

In conclusion, Mr. Chairman, NFPA supports HR 1360 because it recognizes the fire hazards of aboveground storage tanks (AST's), provides incentives to assure that these tanks are properly protected

through the adoption and enforcement of nationally recognized model fire safety codes and allocates a portion of the funds collected under the "Notification" provisions of the bill to enhance code enforcement at the state and local level.

Mr. Chairman, I would be pleased to answer any questions you might have about our position.

<u>APPENDIX</u>

A.)   An NFPA Fire Study; *Three Fire Fighters Die*.....(the
Gadsden, Alabama AST fire of August 31, 1976)

B.)   *Tank Blast Kills Six, Injures Sixty-Four!* (the Kansas City,
Kansas AST fire of August 18, 1959)

C.)   A Review of National Fire Incident Data Bases by NFPA
Fire Analysis and Research Division (1983-1992)

An NFPA Fire Study

Three fire fighters die when exposure hazard ignored

A leased gasoline station just outside the corporate limits of Gadsden, Alabama, became a funeral pyre for three fire fighters last 31 August.

The gasoline station complex, shown in the accompanying diagram, included two 6000-gallon (22.7-m³) aboveground storage tanks, one for regular-grade gasoline and one for premium grade. The storage tanks were set approximately 8 inches (200 mm) off the ground, on concrete saddles.

Both tanks were arranged to feed the dispensing units by gravity. Fuel entered the tanks through a 2-inch (51-mm) piping arrangement that extended from the top of the tank, down the tank head, to near ground level. Each tank was provided with a 1¼-inch (32-mm) atmospheric vent and a second 2½-inch (64-mm) vent with a hinged top that was padlocked closed. The two welded tanks were labeled by a nationally recognized testing labora-

This article is based on a report prepared by John A. Sharry, NFPA Life Safety Specialist, who visited the scene for the NFPA Fire Analysis Department. Mr. Sharry gratefully acknowledges the assistance of Acting Chief Landis D. Glenn and Lt. Jimmy Stewart of the Gadsden Fire Department; Fire Marshal Ray Thornwell and Deputy Fire Marshal W. B. Houston of the Alabama State Fire Marshal's office; Jimmie B. Sutton, Accident Investigation Specialist of U.S. Bureau of Motor Carrier Safety; and Tom Lasseigne, Hazardous Materials Specialist of the National Transportation Safety Board.

tory as suitable for aboveground storage of flammable liquids.

At the time of the fire, tank No. 1 is believed to have been nearly empty of regular gasoline, while tank No. 2 contained about 1200 gallons (4540 liters) of premium-grade gasoline.

Fire breaks out

At about 1:45 P.M. on 31 August, a 10,000-gallon (37.8-m³) tank truck arrived to fill the two aboveground storage tanks. The driver connected his equipment in the usual fashion. Because the tank truck was not equipped with a pump, the driver was provided with an electrically driven pump to fill the aboveground tanks. The pump was labeled by a nationally recognized testing laboratory for use in hazardous locations (Class I, Group D; Class II, Groups F & G); however, the junction box on the pump motor did not have a cover, negating its suitability for hazardous locations. Although the pump was designed for permanent installation, it was used as a portable pump in this situation.

The pump was set midway between the tank truck and storage tank No. 2. A 3-prong, twist-lock plug connected the pump's power-cord to an outdoor electrical receptacle located between tanks No. 1 and No. 2. A 3-inch (76-mm) flexible hose extended from the truck's discharge manifold to the pump, where it was reduced to 2½ inches for attachment to the intake side of the pump. A 2-inch rigid pipe ran from the discharge

side of the pump to the filler line on the tank.

The truck driver had been unloading premium gasoline for about 20 minutes when ignition occurred. The station manager reported first seeing fire near the ground in the area of the pump. The fire quickly became a fireball that enveloped the driver. The driver ran from the area, but the station manager stopped him and helped to extinguish the fire in his clothing. The driver then was taken to a nearby hospital by a passing motorist.

Alarm is sounded

The Gadsden Fire Department received a telephone alarm at 2:20 P.M., and Engine 9, Engine 3, and Fire Medic 1 responded. Arriving fire fighters found the blaze from a large gasoline spill engulfing the rear of the tank truck. They initially attacked with two 1½-inch hose lines supplied by the water in the apparatus tanks.

Shortly after Gadsden fire fighters arrived on the scene, the Hokes Bluff Fire Department and the Glencoe Fire Department arrived on the scene. Each had received a telephone alarm for the fire, which was not within the boundaries of any of the responding departments, but was in the unincorporated Etawah County.

A tanker relay was set up to supply the attack lines operating from the Gadsden pumpers. The total attack consisted of four, and at times five, 1½-inch hose lines. Eventually, a supply line to provide additional water was laid from a hydrant about 1800 feet (550 m) away.

The attack had been underway for about 35 minutes when, at 2:59 P.M., tank No. 2 BLEVEd. The head piece of the tank separated at the weld and traveled 107 feet (33 m) east, striking a truck and a fire department pumper. The remainder of the tank rocketed 240 feet (73 m) through a pine grove, landing in a field. The explosion dislodged tank No. 1 from its saddle, and moved it about 8 feet (2.4 m) from — and at right angles to — its original location.

The BLEVE created a ground-level fireball approximately 150 feet (46 m) in diameter. It mushroomed several hundred feet into the air, igniting combustibles as far away as 600 feet (180 m). The fireball killed the Gadsden fire chief and two fire fighters, and injured at least twenty-eight others. The Gadsden Fire Department treated many minor injuries at the scene, and believes the number of injuries to be much greater than it recorded.

Although the fireball ignited several large fires, including the service station office, six trailers inside an adjacent fenced yard, two mobile homes, numerous vehicles including Gadsden Engine 9, plus several brush fires, the fire fighters who survived the BLEVE quickly brought these fires under control.

What happened?

It appears that the fire originated at or near the electrically driven transfer pump. An initial vapor fire probably burned through the flexible 3-inch hose from the tank truck, resulting in a spill fire constantly fed from the tank truck. This spill fire appears to have been about 20 feet in diameter, and it engulfed the rear of the aluminum-shelled tank truck. Normally, fusible devices on the tank truck would have shut down the fuel flow; however, there was no evidence of fusible devices on the emergency flow controls of this tank truck.

Fire fighting efforts apparently concentrated on the spill fire and the tanker, and the fire exposure to the two storage tanks was ignored until shortly before the BLEVE.

The storage tanks apparently lacked adequate venting. Although venting arrangements should be tested to determine their actual capacity, the amount of venting specified in *NFPA 30 Flammable & Combustible Liquids Code* for an aboveground tank of this size would require a free circular opening of about 6¼ inches (160 mm). The venting provided was only 1¼ inches (32 mm), plus the additional 2½ inches of capped vent. This capped vent, of course, was of little use since it was padlocked.

Why?

A series of human errors appears to have been responsible for the cause of this fire, for its spread, and for the ultimate tragedy.

The fire first appeared in the vicinity of a pump with an exposed electrical junction box. This source could well have ignited the gasoline vapors.

The fire grew to its ultimate size because it was fed by a steady flow of gasoline from the tank truck. A fusible linkage would have stopped this flow, but it appears that no such link was installed.

A second, larger safety vent was provided on the tank that BLEVEd, but someone padlocked it shut. Although the vent probably was undersized, it might have been adequate to prevent the BLEVE. Padlocked shut, it was an invitation to disaster.

Finally, and most importantly, attacking fire fighters concentrated their efforts on extinguishing the fire, and ignored the two gasoline tanks. When tanks of flammable gases or liquids are exposed to fire, it is essential that they be cooled with hose streams to prevent them from BLEVEing.　✛

Further information about fighting fires where the possibility of a BLEVE exists may be found in two recent *Fire Command!* articles: "Two train BLEVEs: different situations require different strategies" (*Fire Command!*, September 1976) includes a list of references. "Volunteers prepared when butane tank explodes" (*Fire Command!*, June 1976) describes a no-attack decision and illustrates the critical importance of training.

Two damaged helmets are mute testimony to the intensity of the fireball. The leather helmet on the right belonged to the Gadsden chief who died in the BLEVE; the polycarbonate helmet on the left was worn by a fire fighter who was injured. The hole in the crown is a burn-through; it was not caused by impact. A polycarbonate helmet worn by one of the slain fire fighters was destroyed in the incident.

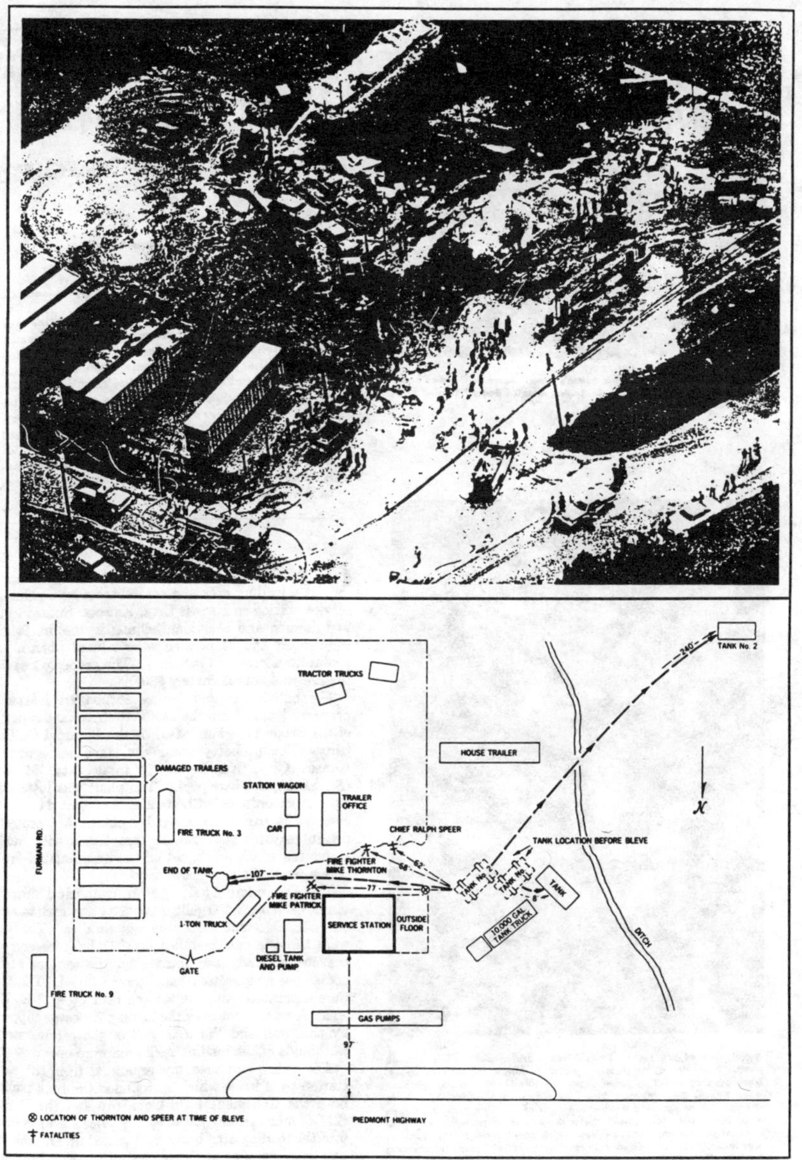

TANK No. 2
240'
TRACTOR TRUCKS
HOUSE TRAILER
DAMAGED TRAILERS
STATION WAGON
TRAILER OFFICE
FIRE TRUCK No. 3
CAR
CHIEF RALPH SPEER
TANK LOCATION BEFORE BLEVE
FURMAN RD.
N
END OF TANK
107'
FIRE FIGHTER MIKE THORNTON
68'
62'
TANK No. 2
TANK No. 1
TANK
FIRE FIGHTER MIKE PATRICK
77'
1-TON TRUCK
SERVICE STATION
OUTSIDE FLOOR
10,000 GAL. TANK TRUCK
DITCH
DIESEL TANK AND PUMP
GATE
FIRE TRUCK No. 9
GAS PUMPS
97'
LOCATION OF THORNTON AND SPEER AT TIME OF BLEVE
FATALITIES
PIEDMONT HIGHWAY

Here is the scene shown on this month's cover. Some of the men in the background were engulfed in the rolling mass of flame. (*United Press International photo.*)

TANK BLAST KILLS SIX,
INJURES SIXTY-FOUR!

by Miles E. Woodworth
NFPA Flammable Liquids Engineer

Another tragedy in the history of fire fighting was recorded in this fire at Kansas City. Millions of people perhaps saw the television news film which captured that terrible moment when a horizontal tank failed and burning gasoline engulfed the fire fighters. Subsequent news reports told how other members of the fire departments at the scene threw themselves on top of the burned victims, or dragged them from the flaming area. The tragedy of this fire, and the lesson it presented, should be memorized by fire fighters and everyone else interested in the safety of fire fighters.

ON AUGUST 18, 1959 at 8:20 A.M. a fire started at a loading rack of a combination bulk plant and service station in Kansas City, Kansas. Subsequently, five firemen and a civilian helping in the fire fighting were killed and 64 firemen were injured when a horizontal tank ruptured violently. The property loss from the fire was approximately $30,000.

The bulk plant and service station are located on property leased from the MKT Railroad and a portion of the property is sub-leased to the Pyramid Oil Company. The property abuts the state line separating Kansas City, Kansas and Kansas City, Missouri. (*See sketch, opposite page.*) The plant was installed in 1927 in accordance with standards of that date. However, as is true of many older plants, fire protection design features were not in compliance with modern standards which are based on the accumulated fire experiences of the past sixty years.

There were four 11 ft. by 30 ft. cylindrical horizontal tanks, each of 21,000 gallons capacity and each mounted on four nine-foot high concrete saddles. The tanks were the originals installed in 1927 but, for corrosion prevention, had been rotated on the saddles and an occasional new section had been installed. The tanks were separated from the loading rack by a 13-in. brick wall. A metal roof over the loading rack was supported by this wall and the rear wall of the service station. Both ends of the loading rack area were open.

The bulk plant area was separated from the service station by a fence which surrounded the bulk plants of both the Continental Oil Company and the Pyramid Oil Company. (*See photo at left.*) Service station windows into the loading area had been blocked up and the fence had been installed at the request of the Fire Prevention Division of the Kansas City, Kansas, Fire Department.

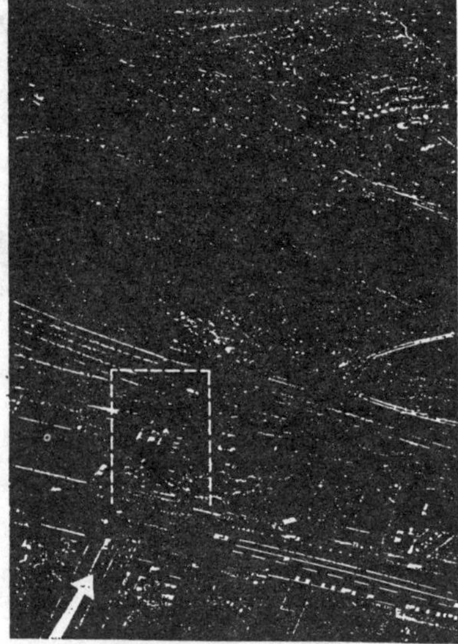

Aerial view of fire scene. Dotted box marks general area of bulk plant. Tank No. 4, at right of group, rocketed forward to street. Dark area of street at lower right was filled with burning gasoline when Kansas City, Missouri, Fire Department responded. With hose streams, the men pushed this burning mass of gasoline back to position in front of tanks, just before the tanks began to rupture. Arrow at lower right indicates position from which cover picture (top picture) was taken. Note burned area in railroad yards at top center, showing how flaming gasoline burst from other horizontal tanks which stayed on their supports. (*Photo by Anderson Photo Company.*)

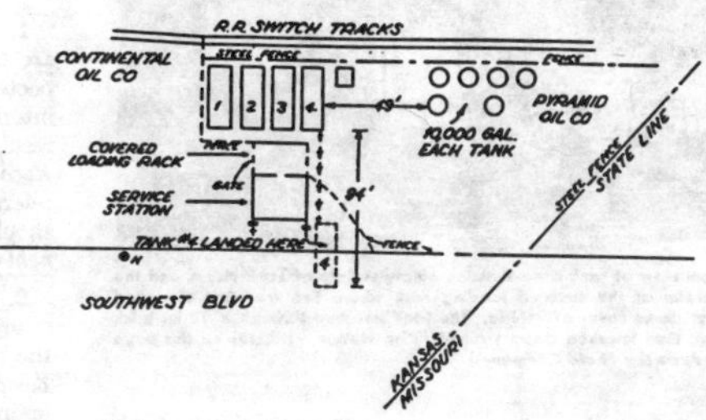

Sketch of bulk plant area showing location of tanks and position of Tank No. 4 after it ruptured. Eruption of burning gasoline followed generally the path of the tank's movement. All during fire fighting action, hose lines cooled the vertical tanks of Pyramid Oil Company at right. The location of this fire at the state line made response from both Kansas City fire departments practically "routine." Actually, the burning gasoline flowed down Southwest Boulevard into Missouri.

Three of the horizontal tanks contained gasoline and one held kerosene. Each tank was equipped with a two-inch combination pressure-vacuum flame arrester type vent. The Pyramid Oil Company tanks were vertical 10,000 gallon tanks and contained both gasoline and fuel oil. A small 275 gallon diesel oil tank was located adjacent to the horizontal tanks. Flow from the horizontal tanks to the loading racks was by gravity. Valves were located in the two-inch piping at the tanks and at the loading rack. In addition, a manual shut-off valve was installed in each fill line for the use of the man loading the tank vehicles.

Events Preceding the Fire

A compartmented tank vehicle was being loaded with two different grades of gasoline with simultaneous delivery being made into two compartments by the driver. Bonding wires between the loading rack and the tank vehicle were in place and the whole system was connected to ground by means of grounding wires. On the previous day Tank No. 1 contained 6,628 gallons of kerosene; Tank No. 2, 15,857 gallons of regular gasoline; Tank No. 3, approximately 3,000 gallons of regular gasoline; and Tank No. 4, 15,655 gallons of premium gasoline. Delivery was being made from Tanks No. 2 and No. 4.

An off-duty company employee had stopped by the plant and was climbing up to the loading platform, which was approximately four feet high constructed of concrete, to show a new unfilled cigarette lighter to the driver. The temperature at 7:50 A.M. was 81 degrees, a south wind was blowing at 12½ miles per hour and humidity was 71 per cent. At 9:50 A.M. the temperature was 85, wind south southwest at 8 miles per hour and the humidity was 66 per cent.

Fire Sequence

The cause of the fire was undetermined. It is known that the off-duty employee was carrying the cigarette lighter in his outstretched hand to show to the driver who was in the process of simultaneously filling two compartments of the tank vehicle. Both men noticed the flame originate at the dome covers but the flash of fire was so rapid that the men were burned and fell off the loading platform. They managed to escape from the area. Only one of the two valves shut off when the driver fell from the platform. As a result, gasoline continued to flow from Tank No. 4 through the open fill line. The resultant fire quickly involved the tanks, loading rack and filling station.

Fire Fighting

The Kansas City, Kansas, Fire Department alarm response was as follows: at 8:20 A.M. the alarm was received; three pumpers, two ladder trucks, and two district chiefs responded. At 8:35 two more pumpers were called; at 8:45 other equipment requested included a specially built deluge truck and foam; and at 9:20 two more pumpers and the off shift were called.

Chief Grass of the Kansas City, Missouri, Fire Department noticed the fire from his office window. After checking with his operator, and knowing that the Kansas City, Kansas, Fire Department was already there, he sent the closest district chief to investigate, because it appeared that the fire was close to the state border.

The district chief upon arrival immediately requested a first alarm assignment of companies. The deputy chief of the Kansas City, Missouri, Department was sent to the fire and he reported burning gasoline flowing down the street about one-half to two-thirds of a block into Missouri. He, therefore, called for additional help.

Although the exact times of the tank failures are unknown (*some time near 10:00 A.M. — Ed.*), it was thought that the overpressure failure of the tanks was progressive from Tanks No. 1 through No. 4. Tank No. 1, holding 6,628 gallons of kerosene, and Tank No. 2, containing 15,857 gallons of gasoline, both failed at their ends facing the railroad tracks. The heads of these tanks tore loose at the weld with some violence, but the tanks moved only about one foot on the concrete saddles. Tank No. 3, containing approximately 3,000 gallons of gasoline, tore loose at the weld only for a short distance near the top of the tank.

Tank No. 4 was the last to fail, approximately two hours after the fire originated. All tanks failed from overpressure, since the under-sized vents were unable to relieve adequately the vapors generated from the boiling liquid.

Table No. 3 of the Flammable Liquids Code, NFPA No. 30, requires a vent for this capacity tank to have a relief capacity of 166,000 cu. ft. of free air per hour. For a tank capable of withstanding five pounds per square inch internal pressure a four-inch free circular opening would be required.

Tank No. 4 rocketed a distance of 94 ft. and landed 15 ft. into the street. En route it went through the 13-in. brick wall separating the loading rack from the tanks and also knocked down one wall of the brick service station building. The end of Tank No. 4 facing the railroad track failed at the moment the tank

Front view of tank area showing empty saddle of Tank No. 4 and the remains of the covered loading rack where fire was first observed near dome cover of vehicle. The tank smashed through a 13-in. brick wall, then knocked down wall of filling station. (Photos on this page by Anderson Photo Company.)

started its skyrocketing action. The other end of the tank came off either en route or after landing. A large ball of fire from the remaining liquid in the tank covered the remaining 85 ft. of the 100 ft. wide street. It was this ball of fire which caught the firemen in its path.

Following the failure of Tank No. 4 at least one of the vertical tanks located at the Pyramid Oil Company was known to have been burning at the vents. Although all the tanks at the Pyramid Oil Company showed signs of fire exposure, they did not appear to be severely damaged.

Eleven fire companies from Kansas City, Missouri were at the fire before Tank No. 4 failed. Following this failure, six companies from the reserve were requested and one of the two off shifts was called back.

The district chief of the first alarm from the Kansas City, Kansas, Department attempted to have the tank valves shut off under the protection of water spray from "fog" nozzles. Due to the intense heat from the fire this was not possible. An attempt was also made to use foam to extinguish or control the fire which was, of course, unsuccessful due to the nature of the fire with its many obstructions and the limited amount of foam which could be applied from any available equipment.

Typical of flammable liquid fires is the problem of burning liquid flowing on the surface of the water used in the fire fighting. In this fire, hose lines covered the front and two sides of the plant and were used to protect exposures as well as being played directly onto the center of the fire. "Fog" nozzles were used to sweep the burning gasoline back to the center of the fire as a part of the procedure to protect exposures. Lines were also used to protect the tanks of the Pyramid Oil Company by keeping them cool. The high temperatures and high humidity made the fire fighting extremely exhausting work.

As reported by a chief officer of the Kansas City, Kansas, Department following the failure of Tank No. 4, those firemen who were uninjured picked themselves up and "fought as if they were mad" going in even closer to the fire than before to fight it.

Five firemen from the Kansas City, Missouri, Department and one civilian spectator, a friend of the men in one of the engine companies, were killed by the skyrocketing of Tank No. 4. Thirty-four firemen from Kansas City, Kansas and 30 firemen from the Kansas City, Missouri Department, and the chiefs of both departments were among those injured.

Conclusions

1. The ends of horizontal flammable liquid tanks are the weakest points. Normally, failure can be expected at the tank ends, either from overpressure or an internal explosion. When one end of a tank fails the resultant jet action will cause a skyrocketing effect. Accordingly, horizontal pressure vessels are sometimes referred to as "bullets" for this reason. So, fire fighters should always keep this basic rule in mind: *NEVER fight any horizontal tank fire from the end!*

2. Cooling water streams should be played directly onto the tanks, if possible. Water so applied will reduce the boiling of the liquid caused by the heat of the fire. Keeping the metal cool above the liquid level also maintains the strength of the steel.

3. Tanks should be equipped with emergency vents to relieve the pressures built up under fire exposures. Every provision for emergency vents contained within Paragraph 2132 of NFPA No. 30, the Flammable Liquids Code, 1959 edition, should be adhered to rigidly. Existing tanks should be brought into compliance with these emergency venting requirements. Inadequate venting is a well proven, distinct hazard to life and property and as such the venting requirements can be made retroactive. *Warning.* Increasing the size of openings in tanks in flammable liquid service may in itself introduce a hazard. For further information see Cleaning of Small Tanks, NFPA No. 327.

4. Locating Class I and Class II flammable liquid loading racks adjacent to tanks and buildings places the potential point of fire origin near these exposures. For this reason the Flammable Liquids Code calls for a 25 foot separation.

5. In this Kansas City incident, the concrete saddles under the tanks remained in place undamaged and again proved their resistance to fire. The hazard of the common practice of using unprotected steel supports is well proven, and this fire unquestionably would have been much more severe if the concrete saddles had not been provided.

Acknowledgments

The writer wishes to express his appreciation to Chief Francis Doherty and Assistant Chief George A. Casey of the Kansas City, Kansas, Fire Department and to Chief Edgar M. Grass of the Kansas City, Missouri, Fire Department for their complete cooperation. Also appreciated is the cooperation of several other individuals who contributed information to the preparation of this report.

View from other end of tanks, near railroad tracks. Note complete failure of Tanks No. 1 and 2, and the slight separation of seam of other tank. All tanks failed from overpressure due to inadequate venting. Perhaps the fact that these tanks remained in position when they failed gave fire fighters a false sense of security in their attack.

Table 3

Required Total Pressure Relief Capacity of Vents

Capacity of Tank		Minimum Total Pressure Relief Capacity (Cu. Ft. of Free Air Per Hour)	Approximate Diameter in Inches of Free Circular Opening for Various Pressures			
Gallons	42-Gallon Barrels		8 In. of Water	1 PSI	2½ PSI	5 PSI
1,000 or less	23.8	25,300	4	2½	2	1½
4,000	95.2	69,500	6¾	3¾	3	2½
18,000	428	139,000	9½	5½	4¼	3¾
25,000	595	166,000	10¼	6	4¾	4

2132. Emergency Relief:

(a) Every aboveground storage tank shall have some form of construction or device that will relieve excessive internal pressure, caused by exposure fires, that might cause the rupture of the tank shell or bottom.

2150. Foundations and Supports: Tanks shall rest directly on the ground or on foundations or supports of concrete, masonry, piling, or steel. Exposed piling or steel supports shall be protected by fire-resistive materials to provide a fire-resistance rating of not less than two hours.

5210. Truck Loading Racks.

5211. Location: Truck loading racks dispensing Class I or Class II flammable liquids shall be separated from tanks, warehouses, other plant buildings, and nearest line of property that may be built upon by a clear distance of not less than 25 feet, measured from the nearest position of any fill stem. Buildings for pumps or for shelter of loading personnel may be part of the loading rack.

Some Comments on NFPA Standard No. 30 — "Flammable Liquids Code"

AFTER the Kansas City fire tragedy, both industry and the fire service are giving more attention to flammable liquid storage facilities. The NFPA Flammable Liquids Code presents basic recommendations which will minimize fire hazards and the fire control problem. The selected paragraphs shown above pertain directly to bulk plant installations.

For fire departments, the Flammable Liquids Code is particularly useful in pre-fire planning. When inspections uncover some situation that does not meet the minimum standards of the Code, then the fire department should immediately modify its fire control planning to be prepared for the worst possible situation.

For example, a tank with inadequate venting is apt to fail if exposed directly to fire. A tank resting on unprotected steel supports can be expected to collapse and perhaps rupture. If loading racks are too close to tanks or buildings, another hazard is presented.

Through the years the Code has been altered as fire experience brought the need for such revision. Here, for example, are some instances which brought such revision.

On October 28, 1954 in Philadelphia, Pa., ten firemen were killed when a pressure tank ruptured violently. A chemical reaction took place in the flammable liquid in the tank and because of a plugged, inadequate sized vent the buildup of pressure from the reaction caused the tank to fail.

On May 18, 1956 two firemen were killed in Galena, Md., when a horizontal tank skyrocketed during a fire when one end of the tank failed. This failure was caused by flame impinging on the tank shell from burning vapors from the vent causing the metal to soften. (Fire Loss Bulletin Series 1956-1.)

On July 19, 1956 nineteen firemen were killed fighting a fire involving a spheroid type tank when it suddenly ruptured releasing a huge ball of burning liquid. This failure also was caused by flame impinging on the tank shell from the burning vapors at the tank vent causing the metal to soften and the sudden rupture of the tank. (NFPA Quarterly reprint Q50-3.)

As a result of these fires Paragraph 2133 was included in the Flammable Liquids Code, NFPA No. 30, (*price 60 cents*) which reads as follows:

"2133. The outlet of all vents and vent drains on tanks designed for 0.5 pounds per square inch or greater pressure shall be arranged to discharge in such a way as to prevent localized overheating of any part of the tank, in the event vapors from such vents are ignited." ....

Storage tanks for flammable liquids are commonly used in three of the property classes identified in the nation's fire incident data bases -- (1) flammable or combustible liquid tank storage (including tank farms and bulk plants), (2) petroleum refinery or natural gas plant, and (3) public service station. With the incident coding now in use, it is not possible to distinguish fires that involve storage tanks from fires that do not, nor to distinguish above-ground tanks from below-ground tanks. Nevertheless, it is possible to make a couple of pertinent observations about the fire experience in these properties where storage tanks are found.

Reported fires in tank storage facilities and refineries are relatively rare, as the table below shows, and the numbers have been declining in recent years. Reported fires in public service stations are much more common than those in tank storage facilities or refineries, but it is likely that few of these involve fire exposure of storage tanks. Only a small fraction of public service station fires begin in the tank area. Reported public service station fires have shown an inconsistent trend, principally downward, in recent years.

Reported Structure Fires in Selected Properties
Having Flammable Liquid Storage Tanks

Year	Tank Storage Facilities	Refineries	Total	Public Service Stations Originated in Product Storage Area (includes storage tanks)
1983	72	125	1,318	73
1984	109	82	1,171	71
1985	94	91	1,424	119
1986	84	71	1,148	45
1987	89	70	1,001	71
1988	107	68	935	88
1989	101	48	840	40
1990	102	58	1,000	42
1991	54	42	1,135	41
1992	75	57	1,198	29

When flammable liquid storage tanks are involved, there is special concern over the dangers to firefighters. National estimates of firefighter injuries cannot be done with the same confidence for the full 10 years of 1983-1992, but 5-year figures for 1988-1992 are possible. During this period, there were an estimated average of 77 reported firefighter injuries per year at public service station structure fires, 27 per year at refinery structure fires, and 17 per year at storage tank facility structure fires. The total of roughly 130 firefighter injuries per year for the three property classes compares to an average of 57,050 firefighter injuries per year at fire scenes at all properties in the same period.

Each of these three property classes also has had two incidents involving fire fighter fatalities in the most recent 10 years for which data is available (1983-1992). The two incidents involving public service stations each involved one fire fighter death due to heart attack at a fire where the victim had not been involved in extended fire fighting and there was no tank involvement. Of the two refinery incidents, one was a single-death incident involving a vehicle accident on the way to the fire, while the other was the 1984 Romeoville incident in which 10 firefighters were killed. Of the two tank storage facility incidents, one was a 1991 North Dakota incident in which several fire fighters, withdrawing from a worsening oil tank fire in their vehicle, lost their way in the smoke and were caught in a flash of flame. Two firefighters died of their injuries. The other incident was a 1984 Mesa, Arizona incident in which one firefighter was killed.

The two 1984 incidents are both discussed at length in NFPA reports which are included in NFPA's preassembled packages (attached) of fire statistics and selected incident narratives on these three property classes. Many diverse lessons are reflected in the reports in these packages, including (1) the importance of compliance with the provisions of NFPA 30; (2) the many parts of the facility that may prove critical to fire development in a particular fire, implying many options for improved fire safety engineering and the importance of a systems approach in considering such options, as is embodied in NFPA 30; and (3) the unique dangers involved in fire fighting in this environment and the need for targeted training of fire fighters in the nature of these dangers and the safe practices to deal with them.

Mr. SWIFT. And now I am happy to recognize Mr. J. L. Tidwell, accompanied by Wayne Senter from Auburn, Washington.

STATEMENT OF J.L. TIDWELL

Mr. TIDWELL. Thank you, Mr. Chairman. I thank you for the opportunity to be here. I am Jim Tidwell, the fire marshal for the city of Fort Worth and the chairman of the Uniform Fire Code Committee of the International Fire Code Institute. The Uniform Fire Code is the most widely used code in the United States, currently adopted in about 35 States.

I bring with me today in my written testimony letters of support for this bill from the International Association of Fire Chiefs, the largest management organization in the fire service in this country, along with support from the International Association of Firefighters, which is the largest union or the largest labor group of the fire service of this country. There is some widespread support in the fire service for this bill.

I am here first and foremost as a firefighter. I have been with the Fort Worth Fire Department for 20 years, the majority of that time on the frontline fighting fires and providing emergency medical service to my community. And it is that experience that has caused me to make this issue a priority.

As you know, and as Mr. O'Neill has already stated, our history in the fire service is rife with disasters that have occurred from aboveground storage tanks being exposed to fires. And typically what would happen after one of these disasters would be that the State and local fire service would mobilize to pass State or local regulations prohibiting such things as unprotected aboveground tanks.

A good example of this was an event that occurred in Kennedale, Texas, just south of Fort Worth in 1968. As an aboveground tank was being filled from a transport truck, it overflowed and was ignited by a spark from the pump on the truck. The ensuing fire belched black smoke and flames that could be seen for several miles.

Responding firefighters, most of whom were volunteers, eventually mounted an aggressive attack on this fire and that was about the time that this unprotected tank failed. The end of the tank blew out from an overpressure, spraying the flaming gasoline across the path of the firefighters and some of the civilians in the area.

The fire chief of Kennedale, and one of his firefighters and a newspaper reporter were killed in that incident, and in addition 28 other civilians and firefighters were injured to varying degrees. This was the incident that caused the Texas Fire Service to mobilize and get State legislation passed basically prohibiting the majority of these aboveground tanks.

Now, in the 1970's and 1980's, of course, we have heard about the leaking underground tanks becoming a serious environmental problem, and in 1988, the Environmental Protection Agency adopted the regulations to address these problems, primarily through leak detection monitoring and remediation of contaminated sites.

I am not here to question the success or the intentions of the efforts to solve the environmental problems of underground tanks. I just need for you to understand that the effect of those regulation, pushing these tanks aboveground, is presenting an enormous problem to the fire service in the United States.

Due to the cost of meeting the regulations for underground tanks, we have seen a distinct move to install those tanks aboveground. The members of the fire service quickly recognized this as a legitimate need. The users of fuel needed a way to avoid expensive and complicated requirements while at the same time protecting the environment. Aboveground tanks seem a logical answer. The problem was how to provide an equivalent level of safety to the public and to firefighters.

A loose knit group of fire service and industry began to work on the issue. The effort grew into a full-blown research and development project with a great deal of input from all the interested parties. The result of that effort is the Uniform Fire Code Appendix II F, and the Uniform Fire Code Standard 79-7. These are the documents that present a safe and sane approach to the storage of flammable and combustible liquids in aboveground tanks. These regulations provide a safe way of accomplishing that mission.

The question keeps arising what does it cost? Let me tell you that you can install an aboveground protected tank with all of the safety requirements in place for approximately the same or less than you can put in an underground storage tank system. We are currently installing 35 of these systems in the city of Fort Worth now. So I have got a fairly good handle of on what it costs.

The difference—probably a good comparison to point out when we talk about protected versus nonprotected tanks, would be maybe a high-rise condominium or an apartment complex where you have fire ratings separating the units. Obviously, the reason for that is so that a neighbor—you will be protected to some extent from a fire in your neighbor's apartment.

I don't think anyone would move into such a facility if it was just built out of plywood and two-by-fours the way single family homes are currently allowed to be constructed. But all of this comes with a price tag.

Another one of the questions that we get asked a lot and I heard it earlier today, it had to do with unfunded mandates. The fire protection section of this bill is not an unfunded mandate. The funding for implementation, as written, is coming from a $50 registration fee. This is a minuscule amount of money when compared to the overall cost of an installation.

A portion of this money would go toward training fire inspectors to implement the provisions of the bill as they relate to the smaller tanks.

How did firefighters get involved with this? Why are we using firefighters to enforce Federal law? For that I want to take a quote from an EPA report, which I happen to agree with. It states: "Local officials such as fire marshals frequently include the inspection of aboveground tanks in the enforcement of local code programs. Such local safety inspections significantly complement the SPCC program and for this reason the training of local officials through that program should be considered." This is what we want to do. We want to be a part of the solution.

Historically, the local fire department has been the lead agency when it comes to regulating these installations and I see no reason to change it. If you like, I would be happy to discuss the Fairfax County incident, where the only government agency that took definitive action to protect that neighborhood in a timely manner was the fire department. Everyone else was taking a long time and some of them never did show up.

I want to close by thanking you for the opportunity to be here today. H.R. 1360 presents a landmark that the fire service can point to Congress as being cognizant of our needs. It is the first piece of environmental legislation that considers our safety and we truly appreciate your consideration. This issue is one of survival for our Nation's firefighters. Thank you.

Mr. SWIFT. Thank you, very much.

[The prepared statement of J.L. Tidwell follows:]

CITY OF FORT WORTH, TEXAS

FORT WORTH FIRE DEPARTMENT
FIRE PREVENTION BUREAU
1000 THROCKMORTON STREET
FORT WORTH, TEXAS 76102
(817) 871-6840

THE HONORABLE AL SWIFT, CHAIRMAN
AND THE U.S. HOUSE OF REPRESENTATIVES
SUBCOMMITTEE ON TRANSPORTATION AND HAZARDOUS MATERIALS

SIRS:

H. R. 1360, "The Safe Aboveground Storage Tank Act", is primarily for the purpose of protecting the environment from releases occurring from aboveground storage tanks. Due to the increasing number of these facilities, and because of the threat they pose to responding firefighters, a portion of the bill encourages certain aboveground storage tanks to be insulated to provide two hour fire protection, along with other nationally recognized safety standards.

Aboveground storage tanks containing flammable liquids have been a nemesis to firefighters. Our history is rife with accounts of firefighter deaths and injuries from fires and explosions involving these devices. The typical scenario is a leak or overfill igniting, exposing a bare steel tank to a high intensity pool fire. The unprotected tank fails quickly, engulfing everything within the path of the escaping burning liquid including firefighters.

In 1988, the U.S. EPA finalized regulations to address leaking underground storage tanks (UST's). These regulations have had an extraordinary impact on the environment, assuring the American public that insidious leaks, which have a history of polluting our drinking water, our recreational waterways, and our soil, will no longer be tolerated. From an environmental standpoint, these regulations have been highly successful.

One of the problems associated with the UST regulations is the expense of compliance. Because of monitoring requirements, retrofit rules, and especially the financial responsibility requirements, many small end users had to decide whether it was fiscally feasible to remain in the fuel business if they had to maintain underground tanks. In addition, larger entities realized that the costs associated with compliance would have a devastating effect on their bottom line.

To deal with these problems, many users began exploring the option of above ground tanks. What they found was a myriad of state and local regulations prohibiting the storage of fuels above ground. The reasons for these regulations can be found in the annals of fire service history. Firefighter deaths and injuries can be found time after time throughout the early to middle 20th century. Each time disaster struck in the form of an aboveground tank explosion resulting in firefighter death and injury, firefighters mobilized to cause state and local regulatory changes prohibiting the practice. By the mid 1960's, fueling from above ground tanks was virtually nonexistent.

By the late 1980's, the fire service realized that political and financial pressures were gaining momentum, and it would be difficult, if not impossible to resist the trend to allow above ground tanks.. Firefighters set out to find a way to allow these systems and still provide an equivalent level of safety. Industry joined in the pursuit to facilitate the acceptance of safe above ground tanks. The outcome of this joint effort was Uniform Fire Code Appendix II F and Uniform Fire Code Standard 79-7. These regulations are based on years of research and development which considered public safety, firefighter safety and environmental concerns. The foundation of the regulations is a requirement for each tank used for dispensing to be protected from fire by a method that will allow a flammable liquid pool fire to burn adjacent to the tank for a period of two hours without catastrophic failure of the tank. It is this requirement that would be codified into federal law if H.R. 1360 were passed. In addition, the tanks would also meet one of the model fire codes for general public and firefighter safety provisions.

Portions of the bill that are significant to the fire service include:

1) It will encourage tank owners to utilize new technology to protect their aboveground tanks from fire exposure, (two hour fire protection), thereby allowing firefighters a "window of opportunity" to address the emergency prior to tank failure;

2) It will provide funding for the purpose of training fire service personnel on inspection criteria for these "protected tanks";

3) It will allow state and local authorities to maintain primary jurisdiction for these tanks.

Please note that the funding mechanism is a $50 registration fee on each tank. This fee is minuscule relative to the cost of any tank installation, protected or otherwise. It will generate enough funds to carry out the training for fire inspectors so they can implement the regulations while they conduct their normal permit inspections of these facilities. Because there is currently a force of several thousand inspectors across the country inspecting these facilities, there is no need to build a new bureaucracy to handle this workload, a workload which will be insignificant if those now inspecting tanks (fire inspectors) are granted primary jurisdiction.

Please remember the words of Thomas Jefferson: *"The care of human life and happiness, and not their destruction, is the first and only legitimate object of good government"*. This bill will, without a doubt, embrace those values. It will save numerous firefighter and civilian lives. It is not "sexy" legislation, therefore it doesn't command the headlines that a crime bill or health legislation will. Passage of this bill is, however, a matter of survival for many firefighters. They and their families will truly benefit, as will the American public if this bill passes. Please carefully consider our nation's firefighters in your deliberations.

Thank you

J. L. Tidwell, Fire Marshal
Chairman, Uniform Fire Code Committee
International Fire Code Institute

INTERNATIONAL ASSOCIATION OF FIRE CHIEFS

4025 Fair Ridge Drive • Fairfax, VA 22033-2868

Telephone: 703-273-0911
FAX: 703-273-9363
ICHIEFS: IAFCHQ

June 14, 1993

The Honorable James P. Moran
U.S. House of Representatives
Washington, DC 20515

Dear Congressman Moran:

The International Association of Fire Chiefs (IAFC) is pleased to support your legislation, HR 1360, to regulate aboveground storage tanks of hazardous substances. Representing 10,000 fire chiefs, and emergency service managers throughout the world, the IAFC is committed to providing citizens with the highest possible level of fire protection while reducing health and safety risks for today's fire/rescue personnel. The Safe Aboveground Storage Tank Act of 1993 address both of these concerns.

As first responders, fire and EMS staff have been endangered and some killed from fires involving substandard aboveground storage tanks. HR 1360 promotes not only fire prevention, but also life safety. It helps protect against fires by ensuring aboveground storage tanks are managed in accordance with Federal standards designed to prevent leaks and spills, and in the event of fire or explosion, provides the nation's fire service with critical information to handle the incident safely and effectively.

The Safe Aboveground Storage Tank Act of 1993 is fire safety legislation which is long overdue. We applaud you for your leadership in sponsoring this legislation. Please let me know what the International Association of Fire Chiefs can do to facilitate the successful passage of HR 1360.

Sincerely,

Chief Gary L. Nichols
President

INTERNATIONAL ASSOCIATION OF FIRE FIGHTERS

ALFRED K. WHITEHEAD
General President

VINCENT J. BOLLON
General Secretary-Treasurer

June 1, 1993

The Honorable Jim Moran
House of Representatives
Washington, DC 20515

Dear Jim:

The International Association of Fire Fighters is pleased to inform you of our strong support for HR 1360, the Safe Aboveground Storage Tank Act of 1993. Increasingly, owners of storage facilities are evading the federal regulations for underground tanks by storing hazardous materials in unregulated aboveground tanks. Regulating aboveground tanks would both prevent dangerous leaks and protect against fire hazards at small storage facilities.

I would like to take this opportunity to suggest an addition to the legislation to further advance the public safety goals of HR 1360. Currently, there is a dire shortage of emergency response personnel who are adequately trained to respond to hazmat incidents involving aboveground storage tanks. To fully address the danger posed by aboveground tanks, the federal government should promote and support specialized training of fire fighters and other local emergency responders.

Specifically, we propose that EPA and/or NIOSH award grants to non-profit organizations for the training of emergency responders who protect facilities with aboveground storage tanks. A grant should also be made available for the development of a specialized curriculum to be used in such training. The grant program could utilize the procedures already in use under SARA Title III.

The IAFF stands ready to work with you on this enhancement to the legislation, and we offer our services in gaining congressional support for this important public safety initiative. On behalf of our organization's 195,000 emergency response personnel, I want to thank you for your commitment to providing a safe and clean environment for all Americans.

Sincerely,

Frederick H. Nesbitt
Director of Governmental Affairs

International Fire Code Institute, Inc.

5360 South Workman Mill Road ♦ Whittier, California 90601 ♦ (310) 699-0124 ♦ Fax (310) 699-8031

March 15, 1993

Congressman Moran
900 Second Street, N.E.
Suite 118
Washington, D.C. 20002

SUBJECT: Safe Aboveground Fuel Storage Act of 1993

Dear Congressman Moran:

The International Fire Code Institute is encouraged by the introduction of the Safe Aboveground Fuel Storage Act of 1993. As written, the proposed legislation appears to provide the environmental and fire protection necessary for regulating aboveground fuel storage tanks. By addressing both environmental and fire protection concerns, the proposed legislation will better safeguard the community and the fire service.

Sincerely yours,

Charles Clawson
Chairman of the Board of Directors

/rss

cc: File

cc0042

National Fire Protection Association

International

Executive Offices
1 Batterymarch Park
P.O. Box 9101
Quincy, Massachusetts 02369-9101 USA
Telephone (617) 770-3000
Telex 200250   Fax (617) 770-0700

Washington Office
Suite 560, 1110 N. Glebe Road
Arlington, VA 22901
Telephone: (703) 516-4546
Fax: (703) 516-4550

March 9, 1993

The Honorable James P. Moran
1523 Longworth House Office Building
Washington, D.C. 20515-4608

Dear Congressman Moran,

We have had the opportunity to work with your staff and provide input to your proposed legislation on ABOVE GROUND STORAGE TANKS. We applaud your efforts to provide federal government support to state and local fire officials who are confronted with the potential dangers of flammable and combustible liquids stored in ABOVE GROUND STORAGE TANKS.

The NFPA technical committees responsible for the national consensus codes and standards applying to flammable and combustible liquids (NFPA 30 and 30A) have also been working to address this problem and we are pleased to note that the legislation references these national consensus codes and standards as minimum standards for state and local enforcement under the exceptions provided for in the legislation.

NFPA is pleased to support your efforts and the proposed legislation. If we can be of further assistance to you or your staff please contact me.

Sincerely,

Anthony R. O'Neill
Vice President, Government Affairs

WSAFC WASHINGTON STATE ASSOCIATION OF FIRE CHIEFS

605 E. 11th, Suite 211 • P.O. Box 7964 • Olympia, Washington 98507-7964 • 206-352-0161 • FAX 206 586-5868

July 8, 1993

Slade Gorton
United States Senate
730 Hart Senate Office Bldg.
Washington, D.C. 20510

Dear Senator Gorton,

Senator Robb and Representative Moran have introduced a bill entitled the "Safe Aboveground Fuel Storage Act of 1993". This bill creates a comprehensive regulatory scheme for the storage of flammable and combustible liquids in aboveground storage tanks.

This proposed legislation permits storage in specially built aboveground tanks designed to withstand external fire exposure for a minimum duration of two hours. Fire experts can attest to the dangers of unprotected aboveground storage tanks that, when exposed to high temperatures, will fail with disastrous results. Because of this hazard, fire codes were adopted to require flammable liquids to be stored in below ground tanks. Now, because of environmental concerns, the trend is to put tanks aboveground once again.

The fire service of the State of Washington recognizes the need for aboveground storage and supports the proposed legislation to allow for such storage in approved two hour protected tanks. We ask that you support this legislation as it moves through the legislative process.

Sincerely,

Otto Jensen,
Administrator

bcc: Lee Wheeler, Chief - Renton

OJ:dr/TANKS93(mw)/7/93

IAFC Supports Arson Prevention Act, Safe Aboveground Storage Tank Act of 1993

The IAFC joined other fire service groups in support of two recent fire-related legislation: the Arson Prevention Act of 1993 and Safe Aboveground Storage Tank Act of 1993.

Applauding the efforts of Senators Richard H. Bryan (D-NV) and Joseph R. Biden (D-DE) and Congressmen Rick Boucher (D-VA) and Jack Brooks (D-TX) for their leadership in introducing the arson legislation, IAFC had letters of support hand delivered to each respective office.

Joined by the International Association of Arson Investigators, the National Fire Protection Association, International Society of Fire Services Instructors, National Volunteer Fire Council, International Association of Fire Fighters, Fire and Emergency Manufacturers and Services Association, and the Volunteer Firemen's Insurance Services, IAFC took part in the strong united force showing commitment to reducing arson throughout the United States.

The act would provide competitive grants to be awarded to as many as ten states, which would assist in providing arson investigation training, boosting new initiatives directed at fraud as a cause of arson, developing new program to combat juvenile arson, improving training resources for rural fire fighter, providing resources for the formation of special arson task forces, and supporting research and new programs directed at civil unrest as a cause of arson.

The IAFC also applauds Congressman James Moran (D-VA) for his leadership in sponsoring HR 1360, the Safe Aboveground Storage Tanks Act. The bill would regulate storage tanks used to store hazardous substances.

As first responders, fire and EMS staff have been endangered and killed from fires involving substandard aboveground storage tanks. HR 1360 promotes not only fire prevention, but also life safety. It helps protect against fires by ensuring aboveground storage tanks are managed in accordance with federal standards designed to prevent leaks and spills, and in the event of fire or explosion, provides the nation's fire service with critical information to handle the incident safely and effectively.

A Senate version of this bill, S. 588, was also introduced into Congress in March.

For a copy of these bills, please contact the IAFC Government Relations Department at ICHIEFS: IAFCGOVT or (703) 273-9815 x. 308 or 309. ✤

New address, programs from International Association of Fire Chiefs Foundation

The International Association of Fire Chiefs Foundation Board of Directors met recently and formulated a plan of reorganization, which included moving the offices out of the Medford Professional Center in York, Pennsylvania, to the new address at 1257 Wiltshire Road, York, PA 17403. The new telephone number is 717-854-9083. With this move the board plans to cut operational expenses and direct more of its resources toward programs and publications.

Applications are still being accepted for the Fire Service Scholarships that are awarded each year in the fall. Direct inquiries to the scholarship committee. There should be 25 of these awards this year.

 **WASHINGTON STATE ASSOCIATION OF FIRE CHIEFS**

605 E. 11th, Suite 211 • P.O. Box 7964 • Olympia, Washington 98507-7964 • 206-352-0161 • FAX 206 586-5868

July 8, 1993

Slade Gorton
United States Senate
730 Hart Senate Office Bldg.
Washington, D.C. 20510

Dear Senator Gorton,

Senator Robb and Representative Moran have introduced a bill entitled the "Safe Aboveground Fuel Storage Act of 1993". This bill creates a comprehensive regulatory scheme for the storage of flammable and combustible liquids in aboveground storage tanks.

This proposed legislation permits storage in specially built aboveground tanks designed to withstand external fire exposure for a minimum duration of two hours. Fire experts can attest to the dangers of unprotected aboveground storage tanks that, when exposed to high temperatures, will fail with disastrous results. Because of this hazard, fire codes were adopted to require flammable liquids to be stored in below ground tanks. Now, because of environmental concerns, the trend is to put tanks aboveground once again.

The fire service of the State of Washington recognizes the need for aboveground storage and supports the proposed legislation to allow for such storage in approved two hour protected tanks. We ask that you support this legislation as it moves through the legislative process.

Sincerely,

Otto Jensen,
Administrator

bcc: Lee Wheeler, Chief - Renton

OJ:dr/TANKS93(mw)/7/93

STATE OF MONTANA
DEPARTMENT OF JUSTICE
LAW ENFORCEMENT SERVICES DIVISION

Joseph P. Mazurek
Attorney General

Scott Hart Building
303 North Roberts, Third Floor
PO Box 201417
Helena, MT 59620-1417
FAX: (406) 444-2759

July 12, 1993

The Honorable Senator Max Baucus
United States Senate
Washington, DC 20510

Administration
(406) 444-3874

Dear Senator Baucus:

Criminal
History
Records
Program
(406) 444-3625

As State Fire Marshal of Montana, I request your support of H.R. 1360, the "Safe Aboveground Storage Tank Act of 1993".

This legislation is the first federal legislation ever proposed, regarding aboveground petroleum storage tanks, which takes into consideration concerns of this nation's fire fighters.

.ninal
Investigation
Bureau
(406) 444-3875

The bill's provisions address specific standards regarding construction, testing and other safety issues which have been of concern to fire officials for years. It also includes methods of secondary containment to be provided to protect human life as well as the environment. These requirements are, in our opinion, of extreme importance in attempting to satisfy our charge of protecting the lives and property of the citizens of the state of Montana.

Fire
Prevention
and
Investigation
Bureau
(406) 444-2050

As a result of many inspections, conducted over the years, at petroleum storage facilities we can attest to the fact there are many owners and operators who have not accepted their responsibility to operate their facility in a safe and conscientious manner. As Ms. Lois Epstein of the EDF has stated, the "**let 'em leak' mentality**" prevails. This attitude must be changed. H.R. 1360 will not only provide needed safety requirements for the public but provide need protection for the environment as well.

Narcotics
Investigation
Bureau
(406) 444-3875

Please support H.R. 1360 for the citizens and the fire service of Montana.

Sincerely,

Bruce Suenram, Chief
Fire Prevention and Investigation Bureau

cc: Dennis Taylor, Deputy Director
     Mike Batista, Administrator
     John Colburn, Executive Director NASFM

**Washington
State
Fire Prevention
Officers**

AL SWIFT
1502 LONGWORTH HOUSE OFF. BLDG.
WASHINGTON D.C. 20515

June 14, 1993

Re: The Safe Aboveground Fuel Storage Act of 1993
     HR 1360   S 588

Dear Representative Swift,

I am the Fire Marshal for the city of Auburn Fire
Department and I have been a firefighter for 14 years. I am
also the Chairman of the Washington State Fire Prevention
Officers Fire Code Committee. Over my career I have fought
many fires and have had to deal with those emergencies that
resulted in multiple deaths due to fire and explosions. Many
businesses have on site aboveground fuel storage tanks (AST).
Some of these are potential bombs which lie in wait to kill
or maim innocent bystanders or our brave firefighters.

Americas fire record, in terms of lives lost and
property destroyed, is worse than just about any other
industrialized nation. This is not a proud banner to wave
and the time to reverse this horrible trend could start
today. One type of small AST known as a fire-protected tank,
provides excellent fire safety protection to the general
public and emergency care providers like firefighters and
police. These tanks are designed to resist the effects of
fire for two hours and prevent a harmful explosion. This
protection gives firefighters an opportunity to safely fight
the building fire.

On Tuesday, March 16, 1993, Senator Robb and
Representative Moran jointly introduced a bill entitled the
"Safe Aboveground Fuel Storage Act of 1993". This bill
creates a comprehensive regulatory scheme for aboveground
storage tanks. As a result of input from the fire-safety
industry, the bill recognizes the effective form of fire and
environmental protection provided by fire-protected tanks and
contains provisions to encourage their use.

Firefighters across the country risk their lives every
day to fight fires and save lives. An unprotected AST at a
fire site is avoidable at a reasonable cost. This bill
provides valuable environmental protection and it promotes
firefighter safety. On behalf of the Washington State Fire
Prevention Officers Code Committee, I urge you to support
"The Safe Aboveground Fuel Storage Act of 1993".

Sincerely,

Wayne Senter, Fire Marshal
1101 "D" N.E. Auburn, WA. 98002
(206) 939-3100

A Division of THE WASHINGTON STATE ASSOCIATION OF FIRE CHIEFS

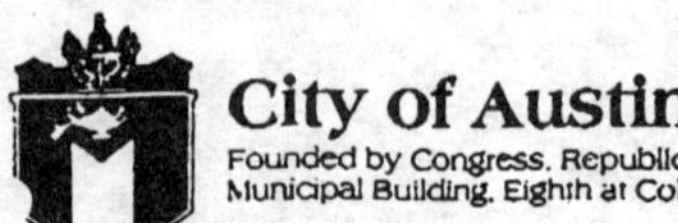

City of Austin

Founded by Congress. Republic of Texas. 1839
Municipal Building. Eighth at Colorado. P.O. Box 1088. Austin. Texas 78767 Telephone 512/499-2000

June 15, 1993

The Honorable Jake Pickle
10th Congressional District
United States House of Representatives
200 East 8th Street
Austin, Texas 78701

Dear Congressman Pickle,

The Austin Fire Department is seeking your support by asking you to vote for the passage of H.R. 1360, known as the *Safe Aboveground Storage Tank Act of 1993*. The proposed legislation will provide environmental protection and protect the safety of firefighters by requiring the aboveground storage of flammable and combustible liquids to be inside of storage tanks which have a high degree of fire-resistance. The impetus of many individuals is to remove underground storage tanks, especially in areas like the Edwards Aquifer and the Barton Creek watershed. The Austin Fire Department has had strong reservations about such actions because most methods of aboveground flammable liquid storage are very vulnerable to fires and large spills. This legislation resolves these problems and provides for enhanced environmental protection. Further, the ability to retroactively apply this legislation to existing sites, where aboveground storage is in less safe, unprotected tanks, will provide public safety organizations with a means of resolving firefighter and public safety threats.

Given the impact this legislation will have to the safety of the public, the Austin Fire Department requests that you support and consider co-sponsoring this bill. Such efforts on your part will surely help expedite the passage of such important bill.

Please feel free to contact me in care of the letter closure if you have any questions concerning the importance of this legislation.

Sincerely,

/s/

Fire Chief Bill Roberts
Austin Fire Department
1621 Festival Beach Road
Austin, Texas 78702
(512) 477-5784

BRR:SAS:sas
XC: The Honorable Curt Weldon, Congressional Fire Service Caucus

FIRE DEPARTMENT

ROBERT S. WILLIAMS
FIRE CHIEF

June 15, 1993

The Honorable Thomas S. Foley, Speaker
United States House of Representatives
1201 Longworth House Office Building
Washington, D.C. 20515

Dear Congressman Foley:

I am Fire Marshal for the City of Spokane. It has recently come to my attention that Congressman Moran and Senator Robb, both of Virginia, have introduced legislation that would regulate above-ground storage tanks. The title of the bill is "Safe Above-Ground Fuel Storage Act of 1993." This bill creates a comprehensive regulatory scheme for above-ground storage tanks and focuses on both the environmental and fire safety concerns associated with the storage of fuel above ground. As you are well aware, this is of particular importance to the City of Spokane and the Spokane area in that we are serviced by a sole source aquifer.

As a result of input from the fire safety industry, the bill recognizes the uniquely effective form of fire and environmental protection provided by fire protected tanks and contains provisions to encourage their use.

I urge you to support the Safe Above-Ground Fuel Storage Act of 1993. Additionally, representatives of a local industry group, including Gordy Lindstrom of the Gelfel Group, wish to meet with you in your office on Thursday, June 17, to speak about the importance of this bill.

I look forward to your support of this endeavor.

Sincerely,

Garry M. Miller
Fire Marshal

cc: Washington State Fire Prevention Officers Assoc.

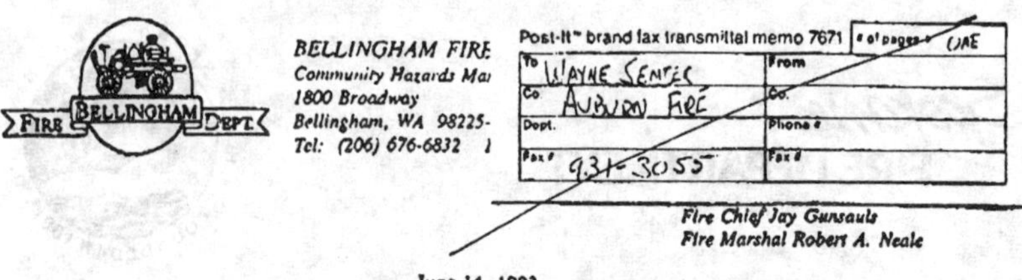

June 14, 1993

The Honorable Congressman Al Swift

1502 Longworth House Office Building
Washington, D. C. 20515

RE: "Safe Above Ground Fuel Storage Act of 1993"

Dear Representative Swift:

I wanted to thank you again for visiting with us recently to hear fire fighters' concerns. I hope the hamburgers suited you!

One issue that has come up time and again is that of above ground storage of flammable and combustible liquids. The very real environmental needs of getting these materials out of the ground is creating a potential public safety problem with which the fire service will have to deal.

The "Safe Above Ground Fuel Storage Act of 1993" sponsored by Rep. Jim Moran and Sen. Charles Robb has addressed many fire service concerns, and follows closely our own Uniform Fire Code requirements. I hope you will support this bill and request a hearing so its benefits can be discussed in a public forum.

If I can be of any assistance discussing the fire safety aspects of storing flammable and combustible liquids, please feel free to contact me at (206) 676-6832. Thank you for your support on this matter.

Sincerely,

Robert A. Neale, Fire Marshal
Community Hazards Management Division
BELLINGHAM FIRE DEPARTMENT

P.S. Thank you for your continued support of the Congressional Fire Caucus and the National Fire Academy, as well.

M E M O R A N D U M

TO: Fire Services Personnel

FROM: J. L. (Jim) Tidwell

DATE: April 14, 1993

RE: Legislative Strategies - Moran/Robb Bills
                     (H.R. 1360/S.B. 588)

Background:

Fire regulatory authorities have been participating with environmental and industry groups in shaping federal legislation which culminated in the introduction on March 16, 1993, of legislation by Congressman Jim Moran (D-Virginia) and Senator Charles Robb (D-Virginia). This proposed legislation provides the first comprehensive federal effort to regulate aboveground storage tanks storing petroleum products.

As introduced, the bill provides for a fee registration program, including inspection for most petroleum storage tanks above 1,100 gallons in capacity. The bill distinguishes between protected tanks (2-hour fire rating) and unprotected tanks, applying different requirements to each. The upgrading to the standards of this bill must be completed within ten (10) years. The bill provides for jurisdiction over these smaller protected tanks to be given to the states rather than EPA. Should the legislation fail to pass, EPA will retain the jurisdictional authority over all of these aboveground storage tanks under the Clean Water Act.

The bill has received support from the Environmental Defense Fund, the NFPA, the International Fire Code Institute, and several state fire service associations. I believe that we also will enjoy very strong support from firefighters' unions. The reasons for this widespread support are indicated below.

With environmental concerns over underground leaking petroleum storage tanks increasing, more fuel is being stored above ground which, if not properly stored, can lead to increased risks of firefighter injuries resulting from exploding petroleum tanks.

Benefits:

1. These bills would assure a level of firefighter safety commensurate with what we have come to expect from underground tanks, i.e., 2-hour fire protection, impact protection, etc. They recognize the environmental and fire safety features of protected tanks and will codify jurisdictional authority over these protected tanks in the states who are expected, in turn, to pass this jurisdiction for inspection to fire regulatory officials since it would be a simple addition to the duties of fire safety officials who are normally permitting these tanks.

2. These bills provide for periodic registration fees for aboveground tanks storing petroleum and for periodic inspection of these tanks. These fees are to be used for the purpose of education and for inspections and are to go to the agencies carrying out these inspections.

3. Retaining jurisdiction in the states and, in turn, the fire regulatory officials is important as it means the fire safety concerns will at least be on equal footing with environmental concerns in these inspections, unlike the existing circumstance where EPA retains jurisdiction and the fire regulatory officials have very little input into the EPA inspection and regulation process.

4. In times of increasingly tight governmental budgets, the fees being generated should provide revenue for inspection and training funds for local fire departments conducting these inspections.

Conclusions:

If we are to continue the downward trend in firefighter injuries and fatalities, we must remain cognizant of the changing environment we work in.

Efforts by other groups can and will impact our mission; further, they may unwittingly place firefighters at greater risk. The net effect of past environmental legislation has been to force many facilities to place their fuels in aboveground tanks. If these tanks are not provided with adequate protection, we may expect a return to the disasters of the 1940's and 50's when multiple firefighter fatalities were commonly caused by the failure of these tanks in fire incidents.

The regulations proposed in H.R. 1360 (Moran) and S.B. 588 (Robb) provide a cost-effective solution to the safety issues associated with aboveground storage tanks.

I would hope that every firefighter, company officer, chief officer, and their respective associations understand that this is an issue of survival. Letters and telephone calls to your congressional representatives can make the difference.

If there are questions or concerns, please contact:

J. L. (Jim) Tidwell
Battalion Chief
Fort Worth Fire Department
1000 Throckmorton Street
Fort Worth, Texas 76102
(817) 871-6808

Mr. SWIFT. Mr. Oxley.

Mr. OXLEY. Thank you, Mr. Chairman.

Mr. DiBona, could you tell me in the real world what is the advantage of a comprehensive standard for terminals and tanks, API's Standard 2610?

Mr. DiBONA. Well, what we are really trying to do here is provide practical, real-world, usable solutions to people who are trying to operate and build these facilities. Many of them are not large companies that have lots of legal and environmental expertise, but often facilities of the kind that have been mentioned here.

And so we believe this document helps those people carry out those jobs. This cites, just to show you the complexity of the thing, 149 different sets of regulations or standards. Fifty-seven of them are API standards. But there are 23 other agencies. One of them is the National Fire Protection Association. There is a chapter in this book that deals with fire protection and prevention.

And this makes it possible for a person who is trying to manage these—it also has the best operating practices. So we think it is a very practical way and it is the first time, incidentally, that such a comprehensive document on this has ever been put together, and we think it is a practical way of making it possible for people who operate these facilities to know what the law is, to find it, to search their way through, and to make sure that they are doing the right thing. That is the reason I think it is incorporated in the law in Florida, in the regulations in Florida. And we hope that it is incorporated wider.

Let me just turn to one other point that Ms. Epstein referred to. A couple of points. I think she misunderstood what I said about our support or opposition of legislation for this bill. We do not think that H.R. 1360 is necessary. We do not think that any new legislation should be passed at this time.

We think that you should let the efforts that are being made by the petroleum industry, the new regulations in many States that are in place, and being written, the current EPA authority, including the SPCC authorities that are being put in place, and you should let those work.

We believe that additional legislation will not be necessary if that happens. But if after all of that activity, the point I was trying to make is if after going through that we find that there are some holes, we would certainly be willing to look at ways of filling those and to work with EPA and the Congress. We don't think that will be the case if you move forward in this way.

Let me say something else about the State authorities. Forty-four States have liability regulations, assigned liability. The people in the petroleum industry are acutely conscious of the potential liability associated with having any kind of spill or contamination.

Mr. OXLEY. Are most of those States strict liability standards?

Mr. DiBONA. Most are strict liability standards. There are some that are strict joint and several. The majority are strict.

Mr. OXLEY. And those are State statutes as opposed to common law?

Mr. DiBONA. I believe it is statute but I am not positive of that. We can give you the citations on all of that.

The point I am making is——

Mr. OXLEY. Let me ask you——

Mr. DIBONA [continuing]. that given your knowledge or the knowledge of the potential liabilities associated with any one of these incidents, one of the strongest defenses you would have, a facility would have or an owner of a facility would have, would be demonstrating that he complied with this standard. And so, therefore, we think that the impetus to use this, even though we cannot as a trade association mandate its use, the States can, and should—but the impetus to use it even in instances where it is not mandated is very strong because it provides a powerful defense to demonstrate that you have used the best practices. So we think it is kind of an important step forward.

Let me finally make one other point about the fire prevention. Most of the accidents that were cited are pretty ancient. They go back to 1950, 1958, and 1976. I mean, we have put in place and are continuing to put in place additional protections. Most of the protected tanks are typically small tanks at service stations, not large AST's at the terminals.

And one of the things that troubled us in this bill that I am surprised the fire protection people did not say anything about, is that the bill would require that all piping be put aboveground, except where infeasible. There are reasons for keeping piping underground that have to do with fire protection, that have to do with access in the case of a fire, that would cause you to keep piping underground, even though it would be feasible to put it aboveground, and we are surprised that they fully support this bill, and don't even comment on a problem like that in the bill.

Mr. OXLEY. Let's ask the gentleman exactly that question. Mr. O'Neill?

Mr. O'NEILL. Yes, we would certainly be concerned with that if, in fact, that was a result of the type of regulations that we are talking about here.

Mr. DIBONA. It is very specific in the bill.

Mr. O'NEILL. In cases where you would have installation where piping would be exposed to damage from vehicles or other types of incidents that would break the piping causing spills or fire.

However, in the larger installation of tank farms there is a lot of piping aboveground now. When you get into the types of incidents that we are most concerned about; namely retail gas stations and that type of thing, you have to run the piping below ground. And the bill would not require that piping to be aboveground. You would have to run it below ground to the dispensing station.

I don't know if Jim wants to add anything to that. He is closer to actual installations than I am.

Mr. TIDWELL. Yes, the portion of the bill that talks about running the piping aboveground, as I understand it, does have to do with tank farms themselves. And that is currently the case in most tank farms, in many tank farms. There is a great deal of piping aboveground. There are ways even at service stations, and some of the industry has done a lot of research in trying to figure out a way to place piping below grade but not underground, and they do this in concrete chases and things like that where they have the containment and they don't have to deal with the fire issue because it is below grade and it is somewhat vaulted and they don't have to deal with the liabilities associated with leaking piping. So there are ways to accomplish what the bill sets out without creating a larger fire safety problem.

Mr. OXLEY. Mr. Mott-Smith, Florida has had a regulatory program since 1983 and a trust fund since 1986, I believe. Do you believe requirement for Federal approval oversight and enforcement will help improve your program? And if so, why?

Mr. MOTT-SMITH. It wouldn't necessarily help our program, but it would help adjacent States and keep, I guess, any contamination coming from their State across State lines into our State, would be one point.

Mr. OXLEY. Have you had problems to that effect?

Mr. MOTT-SMITH. Some minor ones with some smaller facilities.

Mr. OXLEY. Obviously, it would be from the northern part of the State?

Mr. MOTT-SMITH. Alabama and Georgia. I think it would be more helpful just to have nationwide, consistent regulations. If everyone had—if the Federal Government has adopted API Standard 2610, 650, 651, 652, 653, and NFPA 30 and NFPA 30 (a), it would be easier on the regulating community. They would know what they have to comply with and there will be consistency statewide.

Mr. OXLEY. Do you think there may be a difference in application and facilities in Ohio, for example, as opposed to Florida, or do you think there could be a standard national requirement in each and every one of these situations?

Mr. MOTT-SMITH. I think that is where a Federal program could come in. For instance, with the underground storage program now, that is one of the roles that the EPA carries out, providing some consistent standards for the UST program so that they are all pretty much doing things the same way. There are some things that we do differently that we need to do differently in Florida than you need to do differently in Arizona or other States, but in the aboveground program, a similar EPA role would be helpful.

Mr. OXLEY. Would you encourage the States, as Mr. DiBona requested, that they adopt the API Standard 2610?

Mr. MOTT-SMITH. Yes, as well as NFPA 30 (a) and other standards.

Mr. OXLEY. Has Florida done so?

Mr. MOTT-SMITH. Yes, actually on 2610 we are in the process of changing our rule to adopt it. It is a relatively new standard, but we proposed to adopt it and we have not received any opposition.

Mr. DIBONA. The NFPA is in our standard. It covers all kinds of standards. This is an attempt to make it possible for someone to, you know, bring together all of the applicable regulations.

Mr. OXLEY. Let me say, would this take an act by the Florida legislature or the Ohio legislature or could it be enacted by the State EPA?

Mr. MOTT-SMITH. The legislature, of course, gave us the authority to write rules and we have adopted these standards in our rule-making efforts.

Mr. OXLEY. Thank you.

Mr. SWIFT. Thank you.

Mr. DiBona, do you have a number of how many of your member companies' facilities conduct groundwater monitoring?

Mr. DiBona. No, we do not. We do not know how many conduct groundwater.

But I should tell you something, a little bit about the lowest percentage of groundwater monitoring was in the transportation facilities. That is the tank that is associated with transportation facilities. There tends to be a lower level of groundwater to monitoring for a number of reasons. Some of these facilities are quite small. They are 27-barrel tanks. And some of these are just simply used for surge protection in the tanks. That is, it relieves the pressure in the pipelines so the oil is in there sometimes for short periods of time and then reinjected into the pipeline.

Furthermore, pipelines—there is extensive regulation, another set of regulations, that have nothing to do with the aboveground tanks as unique, but affect underground tanks associated with pipelines because they are regulated under another set of Federal laws, and these require extensive cathodic protection. So these long pipelines cross country lines as opposed to the kind of pipelines that we are talking about here, the lines connecting tanks and connecting facilities within a compound, are—that cathodic protection extends to the tanks that are associated with the pipeline.

There is almost no erosion or corrosion of these tanks from the bottom. From the outside of the bottom. To the extent there is corrosion, there is corrosion on the inside of the bottom, not the outside. And it is because of the sulfur or other acidic compounds in the fuel that is being transported through the pipeline and occasionally gets into the tank.

The degree of that corrosion is constantly monitored because there are samples in the tank that are hung in the tank and are frequently measured. So you can tell whether or not the tank is getting thinner on the bottom. So in circumstances like that, and particularly where there are small tanks that are occasionally filled, they have no—they have no monitoring. And there is no reason to believe you should have monitoring there or that it would be a wise expenditure of money because you can quite accurately determine whether there is any existence of any problem for the reasons I have just noted.

For refineries, it is almost universal monitoring. And so it really depends upon the circumstances of the—and the particular surrounding conditions of the particular facility. And what it is used for.

Mr. Swift. What percentage of the industry, do you think, are members of API?

Mr. DiBona. Well, we—we believe that 98 percent of the refineries have monitoring, 80 percent of the marketing terminals and 18 percent of the transportation facilities, and those tend to be the bigger ones which are storing oil and may be even selling oil out of the tankage.

Mr. Swift. That is helpful. What percentage of the whole industry, you know, belongs to API and would, therefore, be responding to your standards?

Mr. DiBona. We certainly have as members of the API something on the order of 80 or 90 percent of the total fuels moved in the United States. Even though there are many small members who—people in the industry who are not members of the API. They don't comprise a large fraction. There are a few large refiners, Coke Oil, and Total, and others who are not members of the API, but by and large, the bulk, and by that I mean 80 to 90 percent range of the total capacity, is in the API.

Mr. SWIFT. Thank you. Ms. Epstein, have you—this is the same question I asked the EPA. Have you done any studies or any estimates on what the benefits of AST regulation would be? There clearly are costs. What are the benefits?

Ms. EPSTEIN. What I did was I looked at the information that EPA developed in its draft liner study and extrapolated from that what the cost of nationwide requirements according to the Moran bill would be, and the benefits, what they would be if you took into consideration the costs of cleanup and health studies and property value depreciation. You find that the benefits are approximately twice the costs. So it is a rough estimate. It is a tough calculation to make.

Mr. SWIFT. Sure.

Ms. EPSTEIN. I was surprised it was that high, and I am pleased. I think it indicates the value of having this type of study. I do want to clarify in your last question, I was looking at the API survey information. It does say that 95 of the 100 refineries across the country are American Petroleum Institute member companies. So there are probably a lot of small companies that aren't members, as Mr. DiBona said. But I think that is—that is an important number, as well.

Mr. SWIFT. Mr. DiBona?

Mr. DiBONA. Yes.

Mr. SWIFT. You wanted to comment?

Mr. DiBONA. About the cost benefit. One aspect of this proposal would be that all existing tanks over, I think, a 10-year period, would be made double-bottom tanks or have release prevention barriers put in. Our estimate of the cost of that is between $17 and $31 billion. And both we and EPA have looked at whether it made economic sense to retrofit, as opposed to putting in release prevention barriers in new tanks, which is the API policy and will be incorporated in the standards in the future.

The conclusion of both EPA and the API is that it would make no economic sense to automatically replace all of the—automatically put release prevention barriers on all existing tankage, aboveground tankage because there are other methods of ensuring the little likelihood of a release, and that is through some of the standards which we have adopted here, which include frequent testing of the tank, some of the expensive things that were mentioned here but less expensive than a release prevention barrier, and the fact that the actual history is, a very small fraction of the problem has come from the tank bottom releases. Or at least in recent years that is the case.

And the problems tend to be in some other areas, which we are also working on, one of which is the releases from buried pipelines, which we are now—which we have now incorporated new standards and methods for testing those into this document. So one of the very specific things asked for in this law is clearly—what would be required in this law is clearly not a thing which we believe passes any kind of risk-benefit test.

Ms. EPSTEIN. I wasn't actually aware that EPA had come to that conclusion yet, since the liner study hadn't been released at this point.

Mr. SWIFT. To be continued.

Ms. EPSTEIN. Right.

Mr. DIBONA. That was their conclusion at the draft study.

Ms. EPSTEIN. Right. Released about 2 years ago with lots of revisions since.

Mr. SWIFT. Mr. Houghton, I didn't want to tar you with some other brush, but as I listen to your testimony, I was reminded of the tobacco industry saying there isn't any evidence and, therefore, you shouldn't do anything. It seems to me that while you are accurate in suggesting that there is a lot we don't know, I would even concede that, therefore, we should be careful what we do.

It seems to me there is plenty of evidence that there is a problem and we should do something. And I didn't—or maybe I wasn't listening carefully enough. I kind of heard you testifying on behalf of your group that, basically, don't do anything. Did I misunderstand you or——

Mr. HOUGHTON. Yes.

Mr. SWIFT. I did?

Mr. HOUGHTON. We would like to see some data that would support what you are doing.

I heard today that in the decade of the 1980's that more product was released by leaking tanks, above-ground tanks, than was in the Valdez incident.

Mr. SWIFT. Yes.

Mr. HOUGHTON. And I am assuming that is correct.

I would like to point out a little thing. Statistics are a very interesting thing. Unless I am sadly mistaken, there—it is a lot easier to clean up a 100-gallon spill in an above-ground tank than it is a 12-million-gallon spill from a tanker. And what I am trying to point out is, this universe is very large and very diverse; and I think the problem with many of these things that you have passed is that, yes, you aim at the major, big problem, but in the meantime, there are all these little busy—businesses out there who don't have fancy lawyers. They don't have excessive engineers. They don't have a lot of payroll. They don't make a lot of money. But they live pretty good in their small towns. They are on the school boards. They are on the town council. They are civic-minded people. They don't go out and deliberately try to contaminate things. But they get hit with the regulations you are passing.

Double-bottom tanks sounds good on above-ground tanks. Excellent. I don't see anything wrong with it except for one thing. Who is going to do it in today's market and stay in business, the little bulk plant with four 18,000-gallon tanks? You are eliminating a vital section of the industry if you are not very careful, if you don't assign your priorities to the risk. You take—every region of the EPA is not the same.

The gentleman was talking about Port Everglades. I am sure that is a major problem. Springfield, Missouri doesn't have the same problem—never will have the same problem, doesn't even have a refinery, doesn't even have a terminal; has a few bulk plants, not as many as they had 20 years ago. That, to me, is my biggest concern with regulation.

The regulations that came out on July 1 have a certification, do you need a response plan or do you not? It is a very simple thing. It is so simple that almost anybody could understand it except for people don't. We have had seven of those certifications mailed to us to ask what to do with them. It says on the form what to do with them. There are a lot of people out there so confused with so much regulation, and they are limited. There are two-, three-, four-person operations. They are limited. They cannot absorb. They are just like a computer, they have only got so much RAM. And, unfortunately, they have reached the limit of their RAM. And you have got to stop, look and think, what is coming out of here? You have got to use reason.

And I am not saying—I have been quoted, and I stand up and I will still stand up; there are many environmental things that are very good. I personally believe the State wouldn't pay for recovery—and the EPA will kill me, and API will kill me for this. I personally think States want to pay for recovery, should be nationwide—bam, no question about it. It is economical, cost effective. Over a period of time especially when gasoline gets up over $1, $1.50 a gallon, which it is going to, it is very economical.

But there are other regulations out there.

John, I am sorry. I don't mean to be 100 percent negative, but I will also add something else. My sister died of cancer from smoking; I definitely am not a tobacco advocate.

Mr. SWIFT. Nor was I suggesting you were. I was suggesting that—

Mr. HOUGHTON. I realize.

Mr. SWIFT. The testimony was spurious. You clarified it for me. Maybe I just didn't get it the first time through. And I have some considerable sympathy for much of what you said in that you are absolutely—I come from an essentially rural congressional district, and I think your comment about we legislate around here for the big and the little, one or two things, they either get swept up into it, which is your case, or they get ignored, which is another thing that occurs sometimes. And those of us from rural areas have to watch out for that all the time.

I am not exactly sure how we do it, but I am sure we have not found a decent way—or strike "decent"; we haven't found an effective way to try and make some distinctions on size and resources that can be brought to bear.

In connection with what I said earlier about trying to find some new models for all of this, if your point is that you and your membership just don't have the same resources as Mr. DiBona and his membership might have, I think that is a point well taken. Whether that suggests you should not have any regulation, I am not sure that is the conclusion. But it does seem to me some risk evaluation, some other things may be useful.

Mr. HOUGHTON. I would like to point out, I have been hearing about the task force today. I would like to know how many PMAA members are on the API task force.

Mr. DiBona comments that 90 percent of the product, they move it. That is fine. And that is probably true. But is that the only movement of the product? How about the 45 percent we sell? How about the 65 percent diesel fuel?

What I am getting at is, over the years, we are ignored many times. We aren't asked. We ask to come over to EPA. We are perfectly willing, and nobody pays our way to come to Washington and—or go to the regulatory agencies. We are very thrilled to be included. I served on the NFPA 30 committee. I know what it is about. I just resigned from it about a year ago.

Things can really get confusing to me. They are talking about a tank farm and a protected tank, and then they talk about the pressure building up, which indicates immediately to me that there wasn't a proper venting. The proper venting has been mandated for, golly, I think since NFPA 30 was written and is now being enforced. I know the State of Missouri went back and retrofitted proper venting on every tank in the State.

I see your point. We have got to get some controls on some of these things. We have got to get it in all levels. But you have got to include all levels in discussion.

Mr. SWIFT. I am very glad we had this, because it clarified for me a lot of what you were saying.

I said before, in this hearing and before, we have got to find a better model than the command and control kind of approach we have used on environmental legislation all these years. That statement is sometimes used as a great Trojan horse in which you drive nonregulation or weaker regulation, and that is not what I am saying. But I just don't think that this model we have got is going to work very much longer. It doesn't get applied fairly. It is unnecessarily expensive, I think, in order the achieve certain policy goals; and I think there are some other models around.

One of the things that you raise that I think needs to be included as we think of how to do this better is, how do you distinguish between the large and the small? It is not that the large is bad.

Mr. HOUGHTON. No, no.

Mr. SWIFT. It is just different. And it is not that the small doesn't create pollution. It has got different resources that it can bring to dealing with it, and we need to work out some way in a new model to allow for all of those things. I think Mr. DiBona and I believe Mr. Tidwell had a comment.

Why don't we go with you, Mr. Tidwell, first?

Mr. TIDWELL. I wanted to respond to the technical issue that was just brought up now regarding venting being a requirement of the NFPA 30, and that is absolutely true. If the emergency venting is in place on one of these tanks, that will prevent the overpressure from happening and it will prevent failure.

The problem is that venting many times is defeated in the field, either at the time of installation or at some time after installation, many times by persons unknowledgeable about what they are actually doing. And when that venting is defeated, then you have built a bomb for responding firefighters to deal with.

So we are not in favor of just providing that one safeguard. There are other safeguards that need to be provided.

Mr. SWIFT. Mr. DiBona?

Mr. DiBona. I would just like to make two quick points. One of them is that—two quick points. One of them is that we did invite PMAA to participate in our committees.

Mr. Houghton. I said the EPA.

Mr. DiBona. You meant EPA. And we had very much in mind the problems of the smaller operator in producing this standard because we realized they have less legal and technical capability and engineering capability. It was intended to make that part of the industry better able to respond to these problems and to—so that all of us are keeping as clean a house as we can with a reasonable effort, and that was the purpose of this publication.

The second point I would like to raise, you did—you were making an earlier statement about the—perhaps excessive law that is in this area, excessive provisions of law; and earlier there was a discussion about the industries trying to work with other parties and put together reasonable ways of going. And there are two things that have happened to the petroleum industry that caused us to be a little skeptical but still want to work with EPA.

One of them was that one of the API members, Amoco Corporation, did agree to—in fact, urged EPA to sit down with it and go through one of its refineries and try and determine how best to regulate this refinery so that you achieved the goals of the various laws but in the most efficient way. And both sets of engineers, EPA's and the Amoco Corporation's engineers, agreed that there was a way of doing that would cost $10 million. But to comply with the way in which you have written the law was $53 million. In the end, they could not do the $10 million job. And there was no difference in the environmental performance, no essential difference in the environmental result. But $53 million was spent and had to be spent. It was a problem associated with the fact that some reasonable—as you said earlier, some reasonable things can't be done.

The second experience that we had is, we did sit down with all the parties on the reformulated gasoline issue. We worked very hard in doing that. We had some—we realized that we were the only people that really had the basic technical information to write the regulation, but we felt that it was more important that we sit down with EPA, with the environmental community, with the farmers and the ethanol people, the automakers, and lay out the technical facts, because we believed in the end we would get a more sensible regulation.

We all worked on that. We all made compromises. And in 1990, I signed the agreement with all of those parties, a solemn agreement, and then that was overturned by EPA on behalf of the ethanol producers. And so it is an attempt, a good-faith attempt to do that by which we are left a bit skeptical.

Mr. Swift. I don't think there is going to be anything easy about getting a new model. Our culture likes adversarial relationships on—more so than any other culture that I am aware of. We are comfortable with being able to duke everything out and we get uncomfortable when we move into a cooperative mode. Regulators are out to keep you from being bad, and as long as they have got that mind-set, they are not—it is going to be very hard for them to move into a mind-set of how do we help them comply? Totally different mind-set.

Mr. DiBONA. I don't think that is right.

Mr. SWIFT. You don't think what is right?

Mr. DiBONA. No, sir. My experience with regulators has not been of that kind. I think they are trying to make the laws you pass work. And I think they have.

Mr. SWIFT. I do, too.

Mr. DiBONA. And I think they have a hell of a time doing it.

Mr. SWIFT. Whether I am right on that or not, it then backs up from them, it seems to me, that the political process, the legislative process and the political input that also sees the world in terms of good guys and bad guys, and we got to go make the bad guys comply.

Mr. DiBONA. Well, I agree with that.

Mr. DiBONA. And in the process, it seems to me, we set up situations in which the model does not seek the least expensive way to meet goals, but rather, it has a—it has this adversarial kind of thing.

Now, it ain't just this side of the dais. You know perfectly well you have got members where if government proposed that the sun came up in the east in the morning, they would be against it.

Mr. DiBONA. I noticed that, too.

Mr. SWIFT. Yes. And you extend this to the environmental movement, you extend it to organized labor.

We have all got strong adversarial ways we bring to this and adopting new models that are more cooperative is going to be tough on all of us. Your troglodytes will say—they will cite the example you just said and say, you cooperate with these people over at EPA and they will nail you for sure, you know. And they are going to have some record to point out.

But EPA can also show points where industry has done outrageous things and has had to really be hit upside the head in order to comply. Not typical, but it has happened, and you can point to it, and it is going to take risk on the part of everybody.

And we have seen some of that go on in how we developed Superfund this year. The environmentalists have got their necks stuck out. Lots of people in industry have got their necks stuck out. And if we fail to pass that thing, everybody who stuck their necks out——

Mr. DiBONA. Will suffer.

Mr. SWIFT [continuing]. is going to in the next Congress, going to point and say, we tried it your way, and we are back to the mattresses again on a major piece of environmental legislation.

So it is not going to be easy. But anybody who looks at the regulatory structure we have on environmental issues in this country today has got to conclude it just can't go on that way, nor can we retreat from the standards. The public's not going to let that happen. So there are all kinds of reasons for people to start taking some risk.

It seems to me what API is doing is a very good thing. Whether it is the whole solution or not, you and I might disagree with that. But the effort you are making which recognizes the problem and addresses it is just excellent, and I think it is another indication that there is some hope we can craft a new model.

And I repeat again, I think Carol Browner isn't given the credit for coming with what is also a very risky thing politically and otherwise, taking on some traditional constituencies in trying to work new models.

I just have one last little set of questions I would like to go over with Mr. Mott-Smith, just—most of these are just for information.

Does Florida's above-ground storage tank program have a fee schedule for tank owners?

Mr. MOTT-SMITH. Yes, it does. There is a $50 initial registration fee, a $25 annual renewal and then there are some minor late fees, things like that. But that is—that generates—again, that is for AST's and UST's. It generates maybe $2 million a year and that goes back into the $160-million fund we have each year for cleaning up contaminated sites.

Mr. SWIFT. Does your program, too, differentiate between types of tanks and different materials that are stored?

Mr. MOTT-SMITH. It does differentiate between the types of tanks. For instance, we have separate regulations for field-erected tanks than we do for shop-fabricated tanks.

Mr. SWIFT. What is the advantage of that?

Mr. MOTT-SMITH. It is pretty much an industry reality. Shop-fabricated tanks are usually less than 50,000 gallons are commonly used in agricultural and retail type of situations. And the field-erected tanks are the larger ones that have to be constructed, the plates welded in the field, and it is the larger bulk storage facilities that have those.

What is stored in them, we just require in our rules that the product stored has to be compatible—rather, the construction materials of the tank have to be compatible with the product that is stored.

We have a lot of rotationally molded polyethylene tanks that are used in the pool and spa industry that we regulate. We regulate mineral acid tanks for storing phosphoric acid, sulfuric acid. They need different types of storage materials, independent of what is in there.

Mr. SWIFT. In your testimony, you said Florida's most costly sites to clean up are the above-ground. Why?

Mr. MOTT-SMITH. It is the—as the gentleman from API suggested, a lot of it is due to historical, but then a lot of it is new, as well. It is just they are bigger tanks, so when you have a release, it is often more catastrophic. You have got large-diameter piping that often has product stored up to 100, 150 PSI, so you have a leak in the joint or a flange or a union, you are going to be losing a lot of fuel in a short amount of time.

Of course, it was common industry practice years ago to drain the consistency off the bottom of the tank, let it flow until they saw product. Or they would use product for mosquito control in the containment area. All of those practices are no longer used anymore, but there are still new and recent incidents of contamination from these facilities.

Back in 1991, a large terminal at Port Everglades lost 20,000 gallons from an overfill. We had a military base lose 4,000 gallons in the last 3 months from one of their tanks. It happens a lot.

But on the other hand, again because of our reliance on groundwater for our drinking water supplies, we decided long ago to go with secondary containment. It was a necessary—it paid dividends recently with Hurricane Andrew. We had a large number of tanks, I think it was over 100, above-ground tanks go down in the storm. But we did not have any incidence of contamination because we, back in 1990, required all our shop-fabricated tanks to have secondary containment. So the containment areas contained the product that was spilled when the tanks blew over in the storm.

Mr. SWIFT. My last question is kind of a reformulation of the one that Mr. Oxley asked you. You have got State programs. They have got a document here and they differ. They are all over the place. That is not to say that they are not good, that there aren't good programs in there.

If we adopt a detailed Federal program, it is going to require lots of changes at the State level. Do you think it is possible to write Federal legislation that establishes goals, standards, if you will, but doesn't necessarily dictate how you get there, that would intrude less into these States' programs than might otherwise occur? And if it is possible, is that a good thing or do you think the uniformity is so important that we should just go ahead and do it at a Federal level and dictate that to the States?

Mr. MOTT-SMITH. I think, again, a good example—and again, I have been with the agency for 17 years, and I have worked in the air program and the hazardous waste program where it was real strict, as you say, command and control. Those were not very pleasant experiences.

I really enjoyed, and have and do enjoy, working with EPA now with their underground storage tank program because that is a program that does provide more or less general guidelines. There are some specific things as well, but you know, pretty much we all think that is a good thing. But it does give the State some flexibility to meet their specific needs.

Again, for instance in Florida, we think second containment is a real good thing. In other places, out west where the ground water is real deep, maybe it is not needed. But there are—I guess there is enough of an umbrella there where we are all pretty much doing everything the same way. And again I think that is important because particularly dealing with the petroleum industry, it is important to be consistent.

As a program manager, I mentioned we contract with county governments to do our inspections. I have got 150 people out there doing inspections every day. And most of the complaints I get are from people that perhaps might be over—exceeding their authority or perhaps being too meek in the implementation of our regulations. And you can't fine, say, a major oil company $1,000 in one county and $100 for the very same violation in another county. I think that same concept, carried over to the national level with some consistency nationwide, would be helpful for AST's. And I think if you were to pattern a Federal AST program like you have the UST program, I think it would be very helpful.

Mr. SWIFT. Thank you.

One last thing, Mr. Houghton. I, too, heard you—I thought I heard you say "API" rather than "EPA" in terms of including the small. And I am glad you said EPA. I am pretty high on that agency right now, but I am going to undertake to ask them if they have—if they have got any input in this process from smaller folks and suggest that if it is not too late in that process to do so, because I think it should be done.

Yes?

Mr. HOUGHTON. Can I make another comment?

Mr. SWIFT. Certainly.

Mr. HOUGHTON. I would like to comment on the regulations of underground tanks. There was a factor that was very important; that was the flexibility in methods of meeting the criteria. In other words, as he mentioned, there is, you know, cathodic protection, inert material for piping and a number of choices, a number of methods of leak detecting. It is a good program and it had good input, and that program did have the PMA input. We can live with it and the State can live with it because it is variable.

I find the biggest problem is when you get into a rigid thing, and that is a problem. Then that is what I mean. I don't see anything wrong with a program down the line that would give a lot of flexibility to each area or—not area but each State to be able to implement it by various methods. And that is something that has to be looked at. There are various methods of doing this.

Mr. SWIFT. Thank you.

Mr. HOUGHTON. Thank you for allowing me to be here.

Mr. SWIFT. I want to thank all of you for being here. It has been extremely helpful. I think this was a good hearing.

Obviously, no legislative action is going to be taken on this legislation this year, but I am hopeful next year people will come back and learn some of the lessons that I think have been taught here today, because it seems to me we have laid the groundwork where, working cooperatively, we can do something that is better than the current system and not onerously so. We probably need to do that.

Thank you all very much. The subcommittee stands adjourned.

[Whereupon, at 12:33 p.m., the hearing was adjourned.]

○